PENGUIN PASSNOTES

Geography

ADVISORY EDITOR: DR S. H. COOTE

GW00597713

Susan Mayhew, B.A., was educated at Leicester University.
She gained a double distinction in her Postgraduate Certificate of
Education at London University. She has taught in a number
of schools in both the public and private sectors and is currently
a tutor in geography at Beechlawn Tutorial College in Oxford.

PENGUIN PASSNOTES

Geography

SUSAN MAYHEW

PENGUIN BOOKS

Penguin Books Ltd, Harmondsworth, Middlesex, England
Penguin Books, 40 West 23rd Street, New York, New York 10010, U.S.A.
Penguin Books Australia Ltd, Ringwood, Victoria, Australia
Penguin Books Canada Ltd, 2801 John Street, Markham, Ontario, Canada L 3R 1 B4
Penguin Books (N.Z.) Ltd, 182–190 Wairau Road, Auckland 10, New Zealand

First published 1984

Made and printed in Great Britain by
Richard Clay (The Chaucer Press) Ltd, Bungay, Suffolk
Filmset in Monophoto Times Roman by
Northumberland Press Ltd, Gateshead

*The publishers are grateful to the following Examination Boards for permission to reproduce
questions from examination papers used in individual titles in the Passnotes series:*

*Associated Examining Board, University of Cambridge Local Examinations Syndicate, Joint
Martriculation Board, University of London School Examinations Department, Oxford and
Cambridge Schools Examination Board, University of Oxford Delegacy of Local Examinations.*

*The Examination Boards accept no responsibility for the accuracy or method of working in
any suggested answers given as models.*

To Jack

Contents

Introduction: How to Use This Book

1. Make sure you know which board and which syllabus you are taking. Each examining board issues a book (usually called *Regulations and syllabuses*) which sets out what you will be required to do and how many marks will be awarded for each section.

2. Check which region you will be studying for the Regional Geography section. Chapter 8 lists the six main options. You will need to study in greater depth your own appropriate section. Many of the themes you will need are outlined in other parts of the book, and you must follow up the page references given.

3. Get some sets of past papers. The addresses of the boards are given at the end of this section. By careful study you will find that each type of paper follows the same order: one question on weather, one on vegetation and so on. You will thus know what type of paper to expect.

4. Work through the whole of this book. Nothing you read in it will be wasted, but there will be areas where you will need more detail than this book can provide. You will need to use your library. Whichever book you are using, use it in conjunction with a good atlas.

5. Practise drawing stylized maps of the countries to help you locate places on a sketch map (fig. 1).

6. Whenever anything reaches the national news, find out where it happened in your atlas. If the news relates in any way to your syllabus, write it in the margin of the appropriate page of this book.

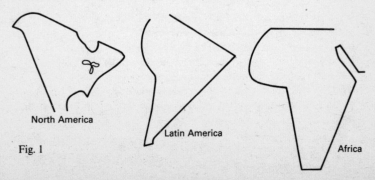

North America

Latin America

Africa

Fig. 1

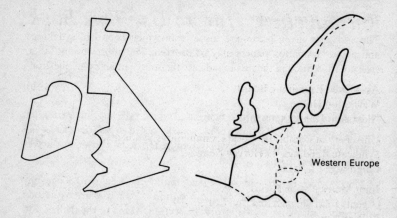

Western Europe

Fig. 1 (cont.)

7. Make a vocabulary book of definitions and learn them carefully.

8. Take with you to the examination:
    two H B pencils
    a good rubber
    a pair of compasses, protractor, set square
    metric ruler
    crayons
Do not waste time using the crayons for 'beautiful' shading. Cross-hatching is just as effective.

9. During the examination:
    (i) Check that you are answering the correct number of questions from the right sections.
    (ii) Obey instructions carefully. Pages 13–14 will explain just what the examiner means by each different command.
    (iii) Use your time sensibly. The first marks in the question are often the easiest ones to get. Make sure you have done something on the full number of questions required.
    (iv) Use any outlines given you. You may mark the map extract, photo-graph or question paper in any way you wish and also make rough notes.

The Examination Boards

The addresses given below are those from which copies of syllabuses and past examination papers may be ordered. The abbreviations (AEB, etc.) are those used in the text to indicate the source of an exam question.

Associated Examining Board, (AEB)
Wellington House,
Aldershot, Hants GU11 1BQ

University of Cambridge Local Examinations Syndicate, (CAM)
Syndicate Buildings, 17 Harvey Road,
Cambridge CB1 2EU

Joint Matriculation Board, (JMB)
(Agent) John Sherratt and Son Ltd,
78 Park Road,
Altrincham, Cheshire WA14 5QQ

University of London School Examinations Department, (LON)
66–72 Gower Street,
London WC1E 6EE

Northern Ireland Schools Examinations Council (NI)
Examinations Office,
Beechill House,
Beechill Road,
Belfast BT8 4RS

Oxford Delegacy of Local Examinations, (OXF)
Ewert Place,
Summertown,
Oxford OX2 7BZ

Oxford and Cambridge Schools Examination Board, (O & C)
10 Trumpington Street,
Cambridge CB2 1QB

Scottish Certificate of Education Examining Board, (SCO)
(Agent) Robert Gibson and Sons, Ltd,
17 Fitzroy Place,
Glasgow G3 7SF

Southern Universities Joint Board, (SUJB)
Cotham Road,
Bristol BS6 6DD

Welsh Joint Education Committee,   (WEL)
245 Western Avenue,
Cardiff CF5 2YX

1. Mapwork: Reading Ordnance Survey Maps

This section is usually compulsory and carries around 15 per cent of all the marks available – sometimes more. If you can 'read' a map you can visualize what the landscape shown on the map looks like. Whenever you can walk over an area looking at its **Ordnance Survey (O.S.)** map you will improve your map reading.

Even good map readers will lose marks, however, if they do not follow the instructions given in the question. In the section below you will find some of the key words used in questions, and examples of how to respond to them. All the **Grid References (G.R.**, see p. 17) in this chapter refer to the map extract of Penzance on the front cover.

Key words in mapwork instructions

1. **Describe:** this includes shapes and sizes. The examiner wants you to use clear, descriptive words, backed up with figures, e.g.

   Q. 1 Describe the shape and size of Ludgvan G.S. (Grid Squares) 5032 and 5033.

   'Ludgvan has a long, narrow, winding shape, as it follows the B3309. It extends along the road for 3/4 km.'

2. **Distribution:** you are expected to say whereabouts a particular feature is found over the map. Don't be vague; refer to areas of the map by:

   (i) their bearing (north, north-east, and so on);

   (ii) their height above sea level ('over 150 m' is better than 'on the high land');

   (iii) their relation to other features, e.g. 'along the course of Trevaylor stream'.

   Q. 2 Describe the distribution of woodland on the map extract.

   'Most of the woodland is between 50 and 150 m, except for the large wooded area in the valley of the Red River (G.S. 5132, 5232).'

3. **State the evidence for:** you are asked to make deductions or back up statements by using evidence FROM THE MAP ONLY. Always give the Grid Reference for the source of your evidence:

Q. 3 Mining has taken place in this area for some time. What is the evidence for this?
'There are spoil heaps in G.S. 4834.'

4. **Explain:** again, use the map evidence to give likely reasons for some feature that occurs on the map.

5. **Relate ... to ...:** you are often asked to link one feature to another, e.g.

Q. 4 Relate the road from Ludgvan to Trevarrack (485318) to the relief.
It is not enough to say 'the road follows the contours'. A better answer would be to say that 'The road rises and falls gently from Ludgvan to 503324 and then is almost level, running parallel to the 20-metre contour.'

6. **Showing your full arithmetical working:** if you have to do a calculation, give the working to show that you know the method used. You will get some marks if the method is right, even if the answer is wrong!

7. **In common:** sometimes the examiner wants to know how certain features are alike. Make this clear in your answer, e.g. '*Like* Trevarrack (4731), Ludgvan (G.S. 5032 and 5033) is also situated where a B road crosses a river.'

8. **Contrasts with; differs from:** stress that you are showing differences by using words like 'whereas' or 'on the other hand'.

9. **Locate:** you could be asked to give a G.R., but you are often asked to draw in a feature on a sketch map you have drawn yourself, or on an outline you have been given. Do this by drawing an arrow to the exact spot and labelling it clearly.

10. **For the area ...:** examiners often want you to use only part of the map. Carefully check where the area they specify is and don't stray outside it.

11. **Comment on:** this means 'locate, describe and explain'.

Often the phrasing of the question will indicate the order in which you should answer it. For example, if a question says '*Describe the physical and human features ...*', you should describe all the physical features before you start on the human ones.

You must know what the examiners mean by the question. Learn the definitions below.

Mapwork vocabulary
Physical geography includes:
   relief: shape and height of the land
   landforms: natural features like hills and cliffs
   drainage: rivers and lakes, marshes
   vegetation.
Human geography includes:
   settlement – isolated buildings, villages, towns.
In this context, note particularly the following four terms:
   site – the piece of ground on which a settlement is built, e.g. hill, river-
     bank
   situation – the location of the settlement in terms of larger features,
     e.g. Guval Downs, the Trevaylor Valley
   form – shape
   function – the services the town provides, e.g. shopping, local govern-
     ment, jobs in industry.
Also included in human geography are:
   communications – tracks, roads, railways, ferries, airports, heliports
   land use – any function of the land.

Mapwork skills

Most mapwork questions set out to test your ability in various fields.
We will look at each in turn.

A. Technical Abilities

You should be able to:
   – recognize map symbols
   – give four-figure grid references
   – give six-figure grid references
   – work out distances
   – work out directions
   – calculate gradients
   – (rarely) draw cross-sections
   – orientate photographs.

Map symbols
The symbols used on the 1:25,000 map differ from those on the 1:50,000.

The symbols on a 1:25,000 ('2½ inch') map are on a larger scale and are not as brightly coloured. The road colourings especially are different.

There are some signs on the 1:50,000 maps which frequently give candidates trouble.

Revision Exercise 1

Using a 1:50,000 O.S. map key (either from a map or a mapwork book) draw the symbols on the same scale for

    (a) paths *and* tracks (these have no dots)

    (b) district *and* county boundaries (these combine dots and dashes)

    (c) electricity transmission line

    (d) pipeline

    (e) cliffs (not covered by tide)

    (f) flat rocks (covered at high tide, coloured blue)

    (g) buildings

    (h) public buildings (public buildings show that a settlement has some kind of **administrative** function, i.e. there is some aspect of running the country going on there)

    (i) cuttings *and* embankments (the 'arrows' always point downslope). See fig. 2.

Fig. 2 *'Cutting' and 'embankment', exaggerated to show arrow points*

OTHER THINGS TO LOOK FOR. Candidates often fail to notice that the width of a beach is shown when high and low water marks are given. The point at which a river flows into the sea is shown when the river bank lines change from blue to black.

Any red dashed lines show paths and tracks that the public has a right to go over (**right of way**); paths and tracks shown in black are not rights of way.

A blue cross can safely be ignored.

Four-figure grid references

O.S. maps are criss-crossed by numbered **grid** lines. The vertical lines are called **eastings** since they number up from west to east, and the horizontal lines are called **northings**. The squares they make are called **grid squares** and you can pinpoint any square, rather as in the game Battleships, by giving the numbers of the grid lines which form its south-west (bottom left) corner (fig. 3). Give the easting before the northing. (Remember, E is before N, alphabetically.)

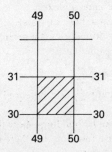

Fig. 3

Six-figure grid references

To pinpoint a spot on a map, a **six-figure grid reference** is used. Firstly the easting is given. For spot X on fig. 3, this is 49. You then estimate how many tenths past the easting grid line your spot is. For example, X is half-way – five tenths – between grid lines 49 and 50, so the first part of the reference is 495.

Next, estimate how many tenths above the northing your spot is. X is about one third of the way between northings 30 and 31; about three tenths. So the second half of the reference is 303. Both sets of three together give the full reference: 495303 (fig. 4).

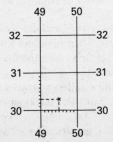

Fig. 4

Here are two ways in which your ability to use six-figure G.R.s might be tested.

Q. 5 With reference to the Penzance extract O.S. map:

   (a) Find Hellangrove Farm (the building, not the words) in square 4834.
   (b) Give a six-figure grid reference for it.
   (c) What is at 505373?

O-level map questions are rarely this simple. Examiners use grid references to draw your attention to a particular point and then ask a question about it. (The answers to parts (b) and (c) are (b) 481342 and (c) a wind pump.)

Scale and distance

The scale of a map is shown by:

1. Its **representative fraction (R.F.).** The R.F. 1:50,000 means that one unit on the map represents 50,000 units on the ground. Thus 1 cm can represent 50,000 cm. 50,000 cm make up half a kilometre. Each side of a grid square is 1 cm, or $\frac{1}{2}$ km long, on this scale map. 1:25,000 maps give much more space to show the landscape, since 1 cm represents 25,000 cm or $\frac{1}{4}$ km.

   Always check the representative fraction on your examination map.
2. The scale diagram (fig. 5).

Fig. 5

You may be asked to measure a distance between two points on the map. Straight lines are easy, but if you measure distances by road you need to follow twists and turns. The quickest way to do this is to use dividers or a compass which does not slip. Set the compass or divider points a convenient distance – perhaps 1 cm – apart, and 'walk' it along the road, turning on each point in turn as you advance (fig. 6).

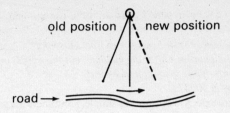

old position   new position

road →

Fig. 6 *'Walking' dividers along a road*

Orientation

Orientation refers to directions. You might be asked:

Q. 6 State the direction of the junction (G.R.) 486333 from the spot height at (G.R.) 494348.

Join the two points using a ruler and a light pencil line. Don't worry about drawing on the O.S. map. The examiner will not see it. Remember to take the bearing from the point specified and express it in terms of the compass rose (fig. 7). On the Penzance extract the junction at 486333 is approximately south-west of spot height 164. The direction asked for may not be absolutely precise.

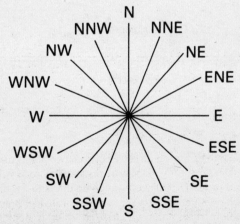

Fig. 7 *A compass rose*

Slightly more difficult to calculate is a **grid bearing**. For this you need a protractor to measure how many degrees round (clockwise) from north your line to the point is (fig. 8).

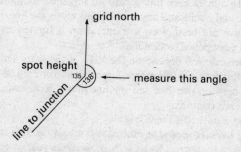

Fig. 8

Gradient

The steepness of the land, or **gradient**, is found by dividing the distance between two points (**horizontal distance**) by the difference in height. First measure the horizontal distance involved (as above), and then find the heights of each point from either spot heights or contours. Subtract the smaller figure from the larger, then divide the answer to this sum into the horizontal distance. Notice that both measurements must be in the same units, i.e. metres.

Q. 7 Showing your full arithmetical working, calculate the average gradient between the two spot heights indicated on the road in grid squares 4734 and 4833.

Distance between spot heights = 1.5 cm, rep. 0.75 km, i.e. 750 m.
Difference between two heights = 143–127 = 16 m.

$$\text{Gradient} = \frac{\text{Distance}}{\text{Height difference}} = \frac{750}{16} = 47 \text{ (to nearest figure)}$$

∴ Gradient = 1 in 47. This is very **gentle**.

Gradients of 1 in 5 or less are **very steep**; between 1 in 5 and 1 in 7 are **steep**. No units are used when gradients are given.

Cross-sections

Occasionally, candidates are asked to draw a cross-section. This is done as follows:

1. Using a ruler, join the two end points you have been given with a pencil line.
2. Fold a piece of paper in half. Lay the folded edge along the pencil line. Make a mark each time a contour hits the line, noting its height. Mark the end points and any spot heights along the line.
3. If you have not been given a frame, draw a framework exactly the length of your pencil section line.
4. Your side scale will depend on the R.F. of the map. Draw side-lines 4 mm apart for 1:25,000 maps, 2 mm apart for 1:50,000 maps. The distance between one line and the line above represents the distance between each contour.
5. Label and number off the side scale.
6. Lay your folded paper along the base-line and, using a ruler, transfer the points to the correct level on the framework (see fig. 9).
7. Join up the points.

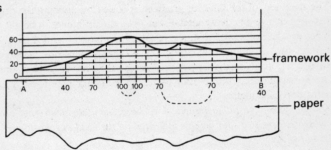

Fig. 9 *Drawing a cross-section*

You should realize that the side scale is not the same as the horizontal scale and that the ups and downs you draw are not in proportion to the real landscape. This distortion is called **vertical exaggeration**.

It might be that a cross-section would help in an answer, although it has not been asked for. A quick **sketch section** can be very useful here. The method is the same as for the cross-section but the measuring, framework and transferring of points can be much more approximate. You could use a sketch section to answer this question:

Q. 8 Compare the valley of Trevaylor stream with that of the stream which meets the sea at Long Rock.

You can see clearly that the Long Rock stream valley is wider and has gentler slopes, whereas Trevaylor stream valley is narrower and steeper-sided (fig. 10).

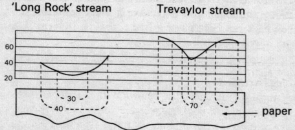

Fig. 10 *Sketch sections*

Photograph orientation

O.S. maps are frequently accompanied by photographs. You can find out which way the camera was pointing by following these three steps:

1. Draw a line down the centre of the photograph.
2. Pick out two features on or near this line; one in the foreground and one further away. Choose the most distinctive features you can: a church with a spire or tower, or a bridge, for example.
3. Find the chosen features on the map. Draw a pencil line to join them up. The line from the nearer to the further feature shows you which way the camera was pointing.

Photographs are frequently used for questions on landforms and land use. With these, and with photo- and mapwork, the more often you see what geographical features look like, on a map and on the ground (**in the field**), the better prepared you will be to tackle the paper. This leads us into the next 'ability' the examiners like to test.

B. *Recognizing geographical features on an O.S. map*

It is not enough, in either physical or human geography, to know the theory: you must know what things look like in the field – or on a photo-graph – and on an O.S. map. Accordingly, when you revise your notes on landforms and settlement, ask yourself the same questions for every feature:

(i) how does it look on a map?

(ii) how does it look in the field?

(iii) can I give an example or location for this feature?

Since all 'mapwork' questions come from British maps, you do not need to be able to recognize features like hot deserts and active volcanoes on a map, although you may have a photograph question on these features.

Revision Exercise 2

Work through the check-list below. With the aid of O.S. maps and possibly a mapwork book* make sure you would recognize the typical contour patterns and/or drainage patterns associated with the following features. Chapter 3 will help you with this exercise.

Rivers/water	Coasts	Glaciation	Upland scenery
V-shaped valley	headland	corrie/cwm/cirque	steep slopes (scarps)
torrent	stack	pyramidal peak	gentle slopes
waterfall	cliffs	arête/ridge	dry valley
interlocking spurs	wave-cut	U-shaped valley	pothole/swallow
incised meander	platform	hanging valley	hole
		truncated spurs	rock drawings/crags
meander	bay	ribbon lake	
flood-plain	beach		
ox-bow lake	spit		
braided river			
estuary			
lake delta			
artificial drainage features			

There is always one best answer in 'recognition' questions; e.g. Little London, 527303, is an isolated rock, but a better answer would identify it as a stack.

* See note on Further Reading, on p. 30

Here is a typical 'recognition' question, based on the Penzance O.S. map extract.

Q. 9 Locate, precisely, one of each of the following: sandy shore, rocky shore, stack, meander, braided stream, V-shaped valley.

Before you attempt this, re-read the notes on the word 'locate' on p. 14.

Frequently, you will be given an outline of part of the map, drawn to scale (fig. 11).

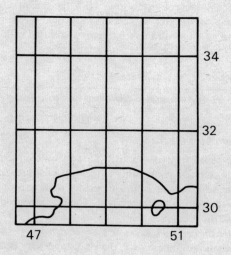

Fig. 11

Q. 9.(cont.) On the outline map supplied, locate and clearly name: a drainage divide (watershed); a spur; a section of river which has been artificially straightened; a valley without a stream.

Do take the trouble to put the information on to the outline with the greatest precision. Mis-located drawings and labellings cost marks. On the other hand, you should use your time in proportion to the number of marks available (shown in square brackets at the right-hand margin of the question paper), so try to strike a balance between being too careless and being too painstaking.

Revision Exercise 3

You should be able to do the same sort of exercise as Exercise 2, this time for **human** (man-made) features:

Settlement	Communications	Land-use
linear village	ring roads	forestry (woods)*
nuclear village	by-passes	orchards*
town centres (CBD)	bus stations*	ornamental parks*
public buildings*	railway stations*	(around 'stately homes')
suburbs	canals	rough grassland*
industrial areas	motorways*	glasshouses*
tourist facilities	motorway junctions*	quarry*
ports and docks	mineral siding*	

* Thorough learning of the O.S. symbols will help with items starred thus*.

C. Description from the map

You may be asked to write about features shown on the map. Sometimes this will combine your ability to identify a feature as in the section above, use the scale to give its measurements, or dimensions, and choose suitable adjectives to indicate its shape. The next question requires you to do all these three things.

Q. 10 Describe the coastal scenery on the map extract.

'Along the whole coastline shown, the land slopes very gently to the sea, with slightly steeper slopes in the west (Penzance) region and the east (Marazion) area. The beach is mostly sand and mud, varying from about 250 m wide in grid squares 4831 and 4830 to 50 m at 473297. There are flat rocks in G.S. 4729 and 4730, each side of Penzance Dock.

'The island of St Michael's Mount is roughly circular, with a small inlet in the south. It rises more steeply from sea level, and may once have been joined to the mainland at Marazion, as it is accessible at low tide. The stacks, e.g. Little London, 527303, and the flat rocks of Great and Little Hogus (513304 and 512307) are also evidence that erosion has been taking place here.'

This answer is *not* something to be learned. It is designed to show you how you can combine geographical terms, geographical directions, place names, descriptive adjectives and precise grid references to make a suitable answer. Don't worry too much about style, though.

USEFUL DESCRIPTIVE TERMS
wide; narrow
gentle; steep
straight; curving, winding
long; short
circular; oval; rectangular; square
high; low

Here is an extract from a report written by the examiners of the London Board. It spells out the difference between a poor answer and a really good one. The candidates have been asked to contrast two towns, Dulverton and Brushford. The marks the examiners awarded are given in brackets.

Poor answer. 'Dulverton is bigger than Brushford [$\frac{1}{2}$].'

Good answer. 'Dulverton is approximately $2\frac{1}{2}$ times as big as Brushford [1] and extends for over 1 km in a curved shape [1], whereas [1, for the comparison] Brushford extends in linear fashion [along a line] for 750 metres [1].'*

Totals for each answer. Poor, $\frac{1}{2}$; good, 4. The good answer scores EIGHT times the total of the poor one!

If you are asked to describe the **distribution** of features (where they are found on the map), look to see if the height of the land is involved: in other words, look for an important contour line. If you study the Penzance extract you will see that the indication of very ancient settlements is above 150 m (461349), and all the present-day villages are below that height.

Being able to pick out an important contour is a very useful skill, especially when describing settlement and land use, particularly farming. (This is because farming is greatly affected by the height of the land and because people today tend to live in the lowlands.)

D. Explanations and deductions

The key to a really good mapwork answer is to show that you understand

---

* University Entrance and School Examinations Council; *Subject Reports, June 1981 G.C.E. Examinations* (*University of London*).

what forces have affected the area shown on the map. Examiners therefore often ask you to explain or work out why the landscape has developed as it has.

WARNING. Don't be too dogmatic. Rather than stating a deduction baldly: 'Long Rock is not very attractive to tourists', show that you are aware that map evidence is only one part of geography. Hence: 'Long Rock *seems* to be less attractive to tourists *because* ...' The questions often reflect this, using terms like '*suggest* reasons' or 'why *might* ... have developed'.

Important note

One or more parts of a mapwork question may ask you to refer to part of the map only. This can be done by referring to the grid lines, e.g. 'west of easting 50 and south of northing 35', or by using a scaled-down diagram (fig. 12). Always confine your answer to the area specified. This is just another way of saying DO WHAT YOU ARE TOLD, one of the most important pieces of advice to an examination candidate.

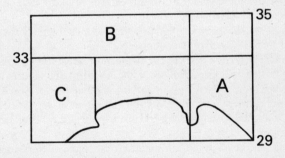

Fig. 12 *Scaled-down map-diagram, showing areas to be referred to*

Revision Exercise 4

When you have reminded yourself of the main influences on settlement and communications by reading Chapter 4, return to these questions, remembering that, in mapwork, you can only refer to factors which are indicated by evidence on the map.

1. Why do so few people live in Castle Gate (493344)? (p. 81)
2. Why is Penzance not a large port? (pp. 114, 119)
3. Give two advantages of Penzance for tourism.
4. What would be the advantages and disadvantages of Long Rock as a holiday site?
5. Quoting from the map, give three ways in which Penzance serves the surrounding area. (p. 85)
6. What map evidence is there that Penzance is the main communication centre of the area? (p. 25)
7. Suggest reasons for the differing routes of the A30 (T) and the railway in the area of east of grid line 50 and south of grid line 34. (pp. 119, 120)

Re-read the instructions on p. 13 (points 3 and 4) before attempting these exercises. The page numbers refer to helpful pages in this book, and will point you towards the right answers, but you must practise for yourself the art of selecting which piece of information in a book is relevant to an O-level answer.

Here is the answer to question 1, as an example to help you. The relevant points are that Castle Gate:

is on a minor route
is on higher ground, i.e. less accessible
has no industry
is in an area of hill farming.

You should add to this by giving the map evidence:

naming the route: B3311
giving the height of the land: 160 m.

You should apply the theories you learn to the evidence provided by the map.

E. Similarities and differences

Questions generally state quite clearly when you are required to find similarities in two features. A sample question might be:

Q. 11 What have the sites of Ludgvan and Trevarrack in common?   [2]

The mark allowance – shown by the number in square brackets – will give an indication of how many points you are expected to note. Two marks would indicate that two simple points are needed. (Think of the person who is marking your paper; think of marks in terms of the 'ticks' he could give you.) In answer to this question, you might mention that Ludgvan and Trevarrack are both located on a valley floor and are both at the bridging point of a stream.

A question requiring differences might use the word 'contrast'. You should re-read point 8 on p. 14, and the 'good answer' on p. 26 before attempting the next question.

Q. 12 Contrast Crowlas (G.S. 5133) with Whitecross (G.S. 5234) by reference to shape, size, site and situation. [10]

The main points which could be given in your answer are noted at the end of this chapter. Notice that there are more than ten (the mark allowance) 'ticks' available, allowing some flexibility in marking.

The word 'compare' is sometimes used. Strictly speaking, this should mean finding similarities *and* differences. Consequently, you must separate the similarities from the differences by putting them into two different paragraphs, clearly indicating which paragraph contains which.

F. Sketch maps

Sometimes you will be asked to draw a sketch map of all or part of the area shown on the map. If you are, follow the direction about scale scrupulously, and only put in what you are told to. At other times you might feel that a sketch map would be the easiest way of putting information over. If you have time, you can do yourself no harm by including a well-labelled sketch map, and you may well gain marks.

Points to remember when drawing sketch maps:
1. If you are not given a scale, choose a simple one – half or quarter that of the O.S. map – and state your scale below the map.
2. Do not draw in every grid line from the O.S. map, but use enough to make your transcribing reasonably accurate.
3. Look for the thickened contours – every fifth contour, – on the 1:50,000 map. These give the broad outlines of relief and help if you are asked to divide the map into relief areas.
4. If you are asked to divide the map into relief (or physical) areas, look for the following four types:
    (i) high and flat (few contours)
    (ii) high and steep, or varied (many contours)
    (iii) low and flat
    (iv) low and varied.

Then use light shading to separate your areas one from another.

5. Spot heights as well as contours indicate relief.

6. A 'sketch map' is not the same as a 'diagram' or a 'section'.

7. The more you practise drawing sketch maps, the easier it will become.

8. Make your labelling accurate and stick to what you are sure of. It is far better to label an upland area 'land over 150 m' than 'chalk upland'. You will know the height of the land from the map; you can only deduce that it is chalk if the word 'chalk' actually appears on the map.

9. Sketch maps are useful in almost every part of the question paper, not just the mapwork section. We will return to this point later.

Finally

Don't think of mapwork as a distinct part of geography. You should look at as many O.S. maps as you can; Chapters 3 and 4 refer you to maps as often as possible. Reading maps is a skill you can learn and can apply to much of the syllabus.

Answer to Question 12

Shape: Crowlas is compact; Whitecross follows the curved line of a minor road running from the A30.

Size: Crowlas is about 250 m north–south and 500 m east–west. Whitecross is much smaller, about 200 m by 50 m. Crowlas is about six times the size of Whitecross.

Site: Crowlas is sited on lower land – about 30 m above sea level on the slopes of the Red River.

Whitecross, however, is located at a height of 70 m on a south-west-facing slope. It lies in a small, dry valley.

Situation: Crowlas is situated on the edge of the coastal plain, about 1 km from the sea, 3 km from the centre of Penzance, on the A30 route from that town.

Whitecross is 1 km further inland. It is on a minor road, which slopes fairly steeply downhill on the way in from Penzance.

Further Reading

The Scenery of Britain in O.S. Maps and Photographs, Book S.2 in E. W. Young, 'Basic Studies in Geography' series (Edward Arnold) will help you, both in this chapter and in Chapter 3.

2. British Isles: Weather and Climate

The earth is surrounded by an **atmosphere** of mixed gases: the air. Processes within the atmosphere make up our day-to-day **weather**; whereas **climate** means a summary of the weather conditions expected over an area throughout the year.

Some boards will give you one 'weather' question each year; others will only base four or five one-mark multiple-choice questions on it. All boards will expect you to know the outline of Britain's climate, its causes and its effects on man's activities.

How weather works

Since this understanding is not deeply examined by all boards, only the bare outlines are given here. Do consult a specialist book if you think you need more detail (see 'Further Reading', p. 44).

Air masses; pressure; fronts

Britain is constantly affected by different **air masses** – bodies of air which vary in their temperature, **humidity** (amount of **water vapour**) and pressure. Different air masses bring different weather conditions. The weight of air pressing upon us is called **air pressure** and is measured in **millibars** (mb). Pressure is marked on weather maps by **isobars** – lines joining places of equal pressure at four- or eight-millibar intervals (figs. 14 and 21).

Pressure system	Air Pressure	Air Properties
Low	under 1012 mb	light and/or rising
High	over 1012 mb	heavy and/or descending

High-pressure systems are called **anticyclones**. On a weather map they have
  – high figures on the isobars
  – isobars spaced far apart.
(i) **Cold anticyclones.** Heavy, cold air brings low temperatures; it *may* also be sunny. We get most of our cold anticyclones in winter.
(ii) **Warm anticyclones.** Air is being forced down from a great height – no one quite knows why. Warm anticyclones bring dry, sunny weather, mainly in our summer, when they can cause heat-waves.
The warm **low-pressure system** affecting Britain is called a **depression**. It involves two air masses: a wedge of warm air trapped within a cold air mass (fig. 13).

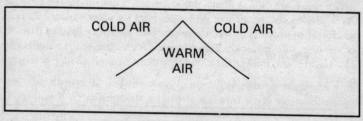

Fig. 13

Depressions come in from the Atlantic, affecting western Britain first and more strongly (i.e. they bring more rain to the west). As they move eastwards, the cold air undercuts the warm air from the back until the warm wedge has been completely lifted up. Gradually the air masses blend. The lines where air masses meet are called **fronts**.
  On a weather map depressions have
    – low figures on the isobars
    – isobars close together
    – a warm and a cold front.
Fig. 14 shows a typical depression and anticyclone side by side.

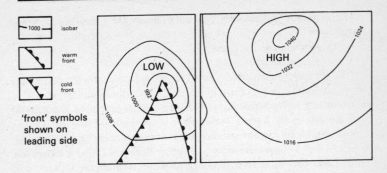

Fig. 14 *Depression and anticyclone to the same scale*

At the fronts parts of the air masses mix (**mixing zone**); cold air cuts under warm air, lifting it up; the warm air is cooled by the cold air. As it gets cooler the air can no longer hold its **water vapour** (water in gas form) so droplets of water **condense out**, gathering around particles of ice, dust, salt or smoke. These droplets form clouds at the fronts (fig. 15). Usually, the thicker the cloud, the more strongly the air is moving upward.

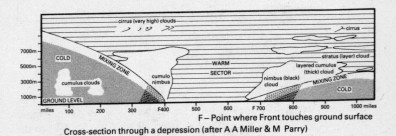

Fig. 15 *Cross-section through a depression*

Use the text and fig. 15, particularly the side scale, to help you with Question 1.

Q. 1 (a) By means of sketches, show the difference between stratus and cumulus clouds.

(b) Which of the two cloud formations would indicate favourable conditions for gliding? Why? [SUJB]

Clouds do not 'burst'. Rain falls when the droplets are heavy enough to overcome the upward-moving air currents. If temperatures are low enough, snow will fall.

Types of rainfall

ALL RAIN RESULTS FROM THE COOLING OF MOIST AIR. The type of rain depends on the type of cooling:

(i) **frontal rain**: warm air cooled by contact with cold air mass (fig. 15)

(ii) **relief rain**: warm air cooled as it rises over uplands (fig. 16a)

(iii) **convection rain**: warm air cooled as it rises with hot air currents (fig. 16b).

(a)

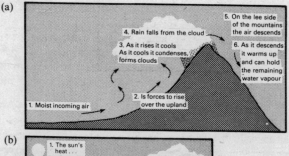

(b)

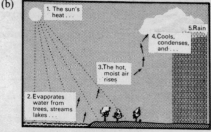

Fig. 16 (*a*) *Relief rain;* (*b*) *Convection rain*

Britain gets mainly frontal and relief rain. Convection rain is less frequent because the weather is not often hot enough to trigger it off. It occurs most often in eastern Britain.

Sometimes the droplets only form **fog**. Again, it is the method of cooling you need to know (fig. 17).

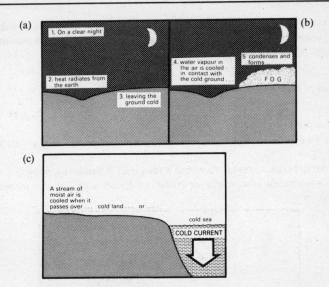

Fig. 17 (*a*) *and* (*b*) *Radiation fog;* (*c*) *Advection fog*

Usually temperatures fall with height above ground, but sometimes heavy, cold air rolls down to a valley floor, undercutting warmer air. This happens in calm, anticyclonic conditions and is called an **inversion**. Fog is often linked with temperature inversions.

Weather maps

1. How the information is collected

Instrument	What It Records
barometer	air pressure
rain gauge	day's **precipitation** (rain, snow, sleet, frost)
weather vane	wind direction
anemometer	wind speed
maximum and minimum thermometer	day's highest and lowest temps.
wet and dry bulb thermometer	air temperature and relative humidity

If you have used the apparatus listed (see table) you will find the next half-question easy to answer, but it is not advisable to tackle it without practical experience. This topic is rarely examined.

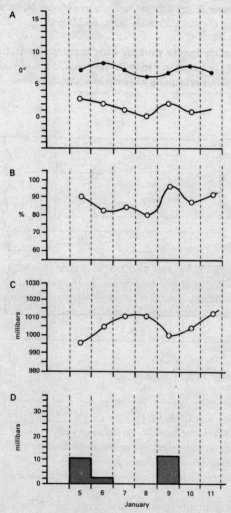

Fig. 18

Q. 2 Using [fig. 18, opposite] (a) Name the instruments which collected the information shown.
   (b) (i) With diagrams describe the instruments used to collect the information on graph A; (ii) Explain how you would read the recordings.
   (c) How were the statistics from graph B obtained?

Questions often involve working with weather graphs, however. The second half of the question (below) is very simple if you read the graphs accurately. Using a set square will help here (fig. 19).

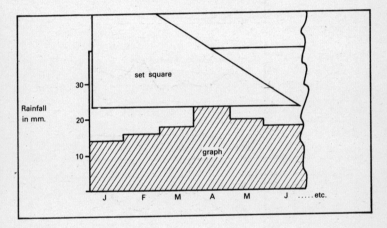

Fig. 19

Q. 2 (cont.) How much rain fell on 5 January?
   What was the relative humidity on 8 January?
   What was the daily temperature range on 10 January?
   Which day experienced the lowest pressure?
   Explain how the passage of a depression would cause the fall shown on graph D?   [CAM, adjusted]
Subtract the minimum from the maximum temperature to get the range; use the appropriate units for each answer.

2. Interpreting weather maps

Weather maps are made daily from readings like those in fig. 20. You should learn the commonest weather symbols, since questions about weather maps frequently occur.

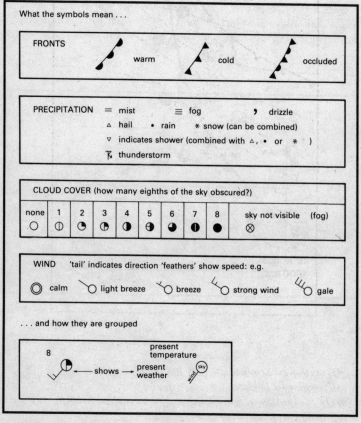

Fig. 20 *Weather symbols*

Question 3 is typical:

Q. 3 (a) Describe and account for the weather shown [fig. 21] in:

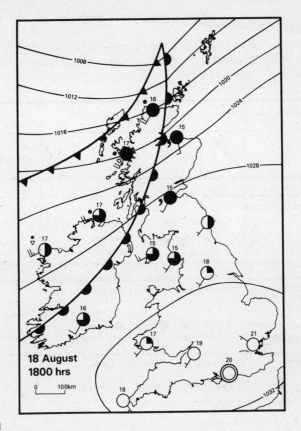

Fig. 21

(i) *north-west Scotland*

(ii) *south-east England*

(b) On . . . *19 August, just after dawn, mist occurred in low-lying areas of south-east England. Account for this.*

(c) On *19 August, rain fell in Manchester but not in London. Why?*

[JMB, adjusted]

You could usefully make a *sketch* section of the warm front *only*, from fig. 14, to answer part (a.i).

Another form of question asks you to work out the weather for yourself.

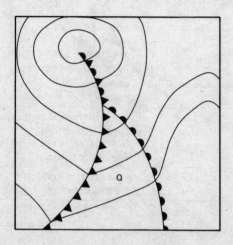

Fig. 22

Q. 4 The weather at Q [fig. 22] could be represented by which of the following?

To find the answer, work out the wind direction by (i) drawing a line parallel to the isobars; (ii) rotating it through 20° towards the centre of the depression. This indicates A or C as the answer. WITH LOW PRESSURE ON YOUR LEFT, THE WIND IS AT YOUR BACK. Remember, Q is in the *warm sector*, so choose accordingly.

Climate

BRITAIN'S CLIMATE: VOCABULARY

Temperatures	**cold**: −5°C to 4°C **cool**: 4°C to 10°C **mild**: 10°C to 15°C **warm**: 15°C upwards
Yearly temperature range	**moderate**: small **extreme**: large
Rainfall	**yearly distribution**: amount each season **heavy**: over 1,500 mm **fairly heavy**: 750 mm–1,500 mm **light**: under 750 mm
General	**mean annual**: average yearly **maritime climate**: mild, heavy rainfall

Factors affecting Britain's climate

1. **Altitude.** Temperatures fall 1°C for each 150 m rise. The isotherms in fig. 24 reflect **sea level** [nought metres O.D.] **temperatures** to eliminate this fall and highlight the effects of . . .
2. **Latitude.** The lower the latitude the greater the heating effect of the sun's rays (see Chapter 7).
3. **The sea.**
   (i) Coastal climates are more moderate, inland climates are more extreme;
   (ii) **Sea breezes** lower temperatures (fig. 23);
   (iii) Warm currents raise temperatures.

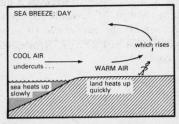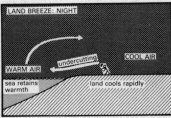

Fig. 23

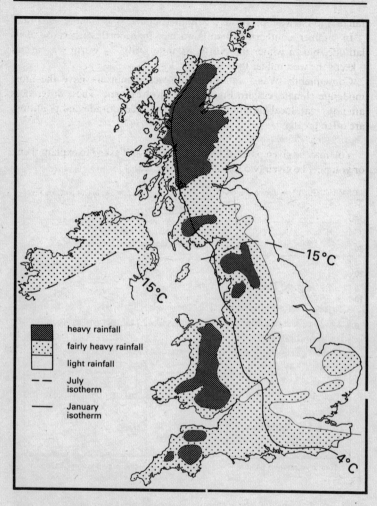

Fig. 24

Britain's climate (see fig. 24)

Rainfall is evenly distributed throughout the year, but is much heavier in the west because (i) depressions from the Atlantic bring frontal rainfall first to the west; (ii) western Britain is higher, so gets more relief rain.

In summer, southern Britain is warmer than northern because of its latitude, but in winter the North Atlantic Drift – a warm sea current – keeps the west milder than the east.

Consequently Wales and the south-west peninsula have the most moderate climate; eastern England the most extreme. These differences are not great: local variations from, for example altitude and pollution, are often greater.

You may be given parts of such a summary and asked to explain them, or you may be given two sets of figures (fig. 25).

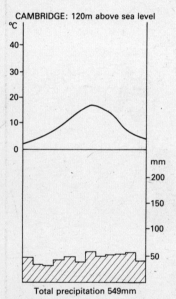

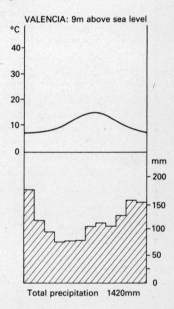

Fig. 25

Q. 5 (*a*) *Account for the difference in mean annual temperature range between the two stations.*

(*b*) (*i*) *Account for the difference in mean annual precipitation between Cambridge and Valencia.*

(*ii*) *Comment on the seasonal distribution of precipitation at both Cambridge and Valencia.*   [LON A, adapted]

To 'account for' or 'comment on' a feature you should first describe it, reading the graph for temperatures and using the 'climate vocabulary' on p. 41 for both temperature and precipitation. Then explain, using the information from this section. Again, you have to know where Cambridge and Valencia are. Look them up. Valencia comes up quite often.

Further Reading
The British Isles, Book 4 in E. W. Young and J. H. Lowry, *A Course in World Geography* (Edward Arnold).

3. British Isles: Landforms

The importance of this section varies very greatly between examining boards. Many candidates learn this whole section very thoroughly even when they will have only one question on it. It is most important that you see some past papers to check on the number of 'landforms' questions to be answered; the syllabus printed by each board will also be helpful. For example, the Joint Matriculation Board do not require you to learn the processes which formed the major landforms, whereas Cambridge wish you to know these processes in detail.

Essay questions on landforms are usually straightforward and involve the following skills:
1. *Recognition* of landforms, by their precise name, from photographs and diagrams.
2. The ability to *describe* the landform, sometimes from a map – this will involve sizes, gradients and shapes.
3. A good *understanding of the processes* involved in forming the feature.

'*Description*' questions may simply use the word 'describe' or the phrase 'what are the characteristics of ...' If you are asked to describe *only*, then restrict yourself to that.

'*Processes*' questions use words like 'account for', 'formation' and, of course, 'process'. Far too many candidates fail to tackle this type of question properly. There is usually a precise word for each process concerned. A candidate who talks about ice 'bull-dozing' or 'sandpapering' will get about half a mark; 'plucking' and 'abrasion' are the correct terms here and will gain the full two marks.

Credit is generally given to students who show they have seen these things for themselves. If you have seen a corrie you might use wording like: 'The corrie we saw at Llyn Idwal had very steep sides, rising to 650 m ...'

Several boards now use multiple-choice questions to test knowledge of landforms. To answer these you *must* know the definition of each geographical term as you will have to choose from a number of likely sounding alternatives. (See Question 4 on p. 52.)

Many questions mix landforms with human geography, asking you to recognize and account for the formation of a landform in the first part and to comment on its effect on settlement and land use in the second. You must be careful not to compartmentalize your knowledge in separate boxes, but should be able to tie up landforms with human geography.

A final warning against thinking too rigidly. In a recent O-level paper, the photograph showed a glaciated landscape together with the river features which had developed after glaciation. Many people failed to recognize a simple V-shaped river valley because they thought it could not exist in a glaciated upland. (See point X, fig. 65.)

Main relief features of Britain

All examiners want you to know where different landforms occur in the British Isles, and the next exercise aims to help your general knowledge.

Revision Exercise 1
1. Trace the outline of Britain – selected geology (fig. 26). Write in the names of the physical features shown on the map, using the letters A–S and add the name of the rock from which they are formed:
   Thus: H = Cotswold Hills – limestone.
2. Make a table as shown below. Fill in the place names you found out from question 1.

Granite Uplands	Volcanic Uplands	Limestone Uplands	Chalk Uplands	Uplands Made of Several Sedimentary Rocks

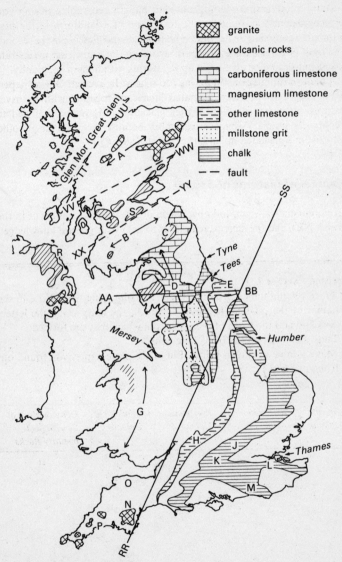

Fig. 26 *Britain – selected geology*

3. Mark in on your map the following lowlands: Vale of York; The Fens; Vale of Evesham.

Examiners may test this knowledge directly, as in the following:

Q. 1 Name TWO limestone uplands, THREE chalk uplands and TWO lowland vales.

The question then goes on to ask you to construct a sketch section along a given line. If you practise a few of these, they will not be so unnerving.

Sketch sections

1. Drawing sketch sections from an atlas

As you did in the mapwork section (p. 21) put the folded edge of a piece of paper along the line the examiners suggest (fig. 27)

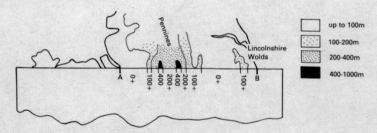

Fig. 27

and continue as for mapwork cross-sections (see fig. 28).

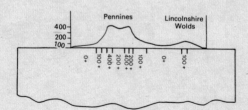

Fig. 28

You will see that no framework lines or construction lines have been drawn, as this is a sketch section. You will note that limestone uplands are generally higher than chalk uplands.

2. Drawing sketch sections from a geology map (e.g. fig.26)
Bearing the last sentence in mind, and the fact that the map shows the upland-forming hard (**resistant**) rocks, attempt a cross-section along the line AA–BB. Use the same process as in 1.

Answering questions on sketch sections
Alternatively you may be given a sketch section. Here are two multiple-choice questions based on the section in fig. 29, which extends from the estuary of the River Tees to the Humber estuary:

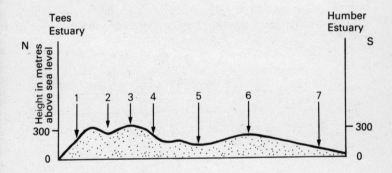

Fig. 29

Q. 2 The upland at 2 is the
 A. Northern Pennines
 B. Peak District
 C. North York Moors
 D. Yorkshire Wolds
 E. Lincolnshire Wolds

Q. 3 The underlying rock at 6 is
  A. chalk
  B. sandstone
  C. millstone grit
  D. limestone
  E. shale
Remember that the section runs north to south in this case.

Highland and Lowland Britain

The line RR–SS across fig. 26 divides Britain very roughly into highland (to the north and west of the line) and lowland (south and east of the line) Britain.

Revision Exercise 2

With the aid of an atlas and fig. 26, complete the table below, which will then summarize the main contrasts between these two divisions of Britain.

	Highland Britain	*Lowland Britain*
Main rocks		
Average height of land (above how many metres?)		
Rivers: long or short (give examples)		

Faulting

A **fault** is a line of fracture through a rock.

You will notice three fault lines on the map (fig 26); VV–WW, XX–YY and TT–UU. In the case of the first two, the land between the faults has fallen, making Central Scotland a **rift valley**; Glen Mor (the Great Glen) is a **fault-guided valley** (fig. 30).

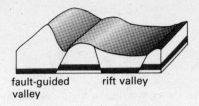

fault-guided   rift valley
valley

Fig. 30 *Faulting and valley formation*

Rocks and landscapes

The rocks in fig. 26 fall into two types:

(i) **Igneous**: made from molten magma welling up inside the earth. **Volcanic rocks** actually came to the surface and spread over it, but **intrusive** rocks like **granite** cooled and solidified underground.

(ii) **Sedimentary**: made by rock waste, or the skeletons of tiny animals, being deposited in layers, usually under water. The rock is then usually compressed and becomes harder. Chalk, limestones and millstone grit are examples of sedimentary rocks. Slates, shales, clays and sandstones are also in this group.

These rock types give rise to different types of scenery.

1. Volcanic rocks

These are usually hard, and can form uplands – like the ones in the Central Valley of Scotland – or small hills, where the **plug** of lava that once occupied the crater is left after the rest of the cone has been eroded away. Edinburgh Castle is built on a volcanic plug.

When the lava from the volcano was more runny it covered large areas and formed **plateaux** (areas of flat, usually high, land). You will see that there was one such large plateau formed in Northern Ireland (fig. 26), the Antrim Plateau.

2. Granite scenery

Granite forms uplands, like the **moors** you identified from fig. 26. Often they have features called **tors**, which look like man-made boulder piles, but these are simply the result of natural **jointing** of the granite as it cools (fig. 31 overleaf).

Fig. 31 *Part of Coombestone Tor, Devon*

3. Chalk scenery

Chalk is not particularly resistant, but it forms uplands because rain-water generally soaks through this **porous** rock (a rock with little pores or holes, through which water can travel). If there is little water flowing over the surface, then very little of the rock will be **eroded**, or removed.

Inland, therefore, chalk forms uplands with one steep slope or **scarp**, and a very gentle downwards slope on the other side (**dip-slope**).

Most of the water in chalk soaks underground until it meets an **impermeable** rock (one which will not let water through). The water is then stored in the porous rock (any rock which will store water is called an **aquifer**).

The top of this saturated layer is called the **water table**. An examiner might define it slightly differently, as:

Q. 4 'The level below which a rock is saturated with water': This is a definition of a (an)

  A. aquifer

  B. artesian basin

  C. spring line

  D. watershed

  E. water table   [LON B]

Obviously, it is not enough to learn your definitions parrot-fashion. You must understand them so that you will recognize them in a slightly altered form.

When the water table is at ground level, water will flow out through fissures in the chalk. The level of the water table is higher in rainy weather and so some springs only occur in the wetter season (usually winter).

When springs occur along the chalk in this way, they are said to form a **spring line**.

Many chalk areas have valleys which are not occupied by rivers, and those are known as **dry valleys**. No one knows exactly how they formed. Here are two possible explanations:

1. The climate was wetter in past times, so the water table, and thus the spring line, was higher. As the climate became drier, the water table fell, and the rivers 'disappeared'.
2. The valleys were cut by the melted water from glaciers. The dry valley in fig. 32 twists from 355847 to 377860.

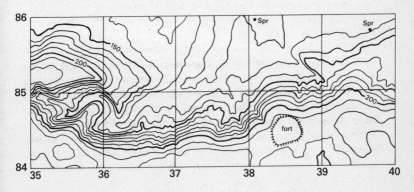

Fig. 32 *Part of Lambourn Downs scarp*

Revision Exercise 3

1. Draw a sketch section along grid line 39 (fig. 32) and label 'scarp slope' and 'dip slope'. Use a side (vertical) scale of 2 mm for each contour interval.
2. Give the dimensions – length, width, height – of the dry valley from 355847 to 377860. Use the R.F. to tell you the scale.

4. Limestone scenery

All limestones form uplands, but Carboniferous Limestone gives rise to a very characteristic scenery. It is the result of a weak acid (carbon dioxide, from the air, dissolved in rainwater) attacking the joints down and along

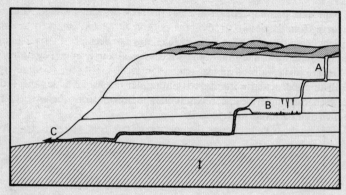

Fig. 33

the rock. Fig. 33 summarizes the features of limestone scenery. It shows a stream disappearing down a **pothole** at A. At B it has widened out a **cave** (B) where **stalagmites** (ground) and **stalactites** (ceiling) form as the water evaporates. If the cave roof should collapse, a **gorge** might form. When the limestone gives way to an impermeable rock (I) the water will come out as a spring (C). The bare limestone at D is called a **pavement**.

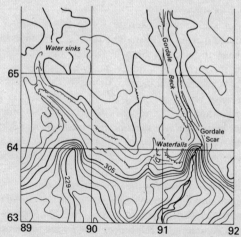

Fig. 34 *Limestone scenery (extract from O.S. 1:50,000, Malham area)*

Notice on the map extract (fig. 34) the rock drawings which indicate a gorge from 913651 to 915642, the pothole at 895655 (marked 'water springs') and the dry valley from 893653 down to 900650.

Carboniferous Limestone scenery is known as **karst**.

Underground water is an important source of supply for many areas in Britain, particularly London. If an aquifer (see p. 52) is sandwiched between two impermeable rocks it acts as an underground reservoir known as an **artesian well** (fig. 35).

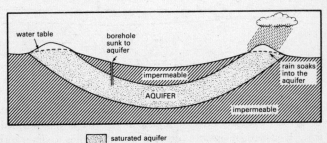

Fig. 35 *An artesian basin*

Like most questions on landforms, Question 5 is in many parts:

Q. 5 (a) Draw only a fully labelled diagram to explain the formation of a scarp foot spring.

Use the section you drew in Exercise 3, 1 (p. 53) to help you with this. Copy the outline of your section and add a layer of impermeable rocks as in fig. 33 beneath the chalk. Show the water table as in fig. 35, running a little below the surface as a dotted line which comes to ground level where the chalk meets the impermeable rock. It is here that the spring will emerge.

The question continues:

Q. 5 (b) Name one area where such springs occur [see fig. 32].

  (c) Explain the process by which ground water acts on Carboniferous Limestone and name the surface features which result from this action.

You will need a textbook for the second part of this.

  (d) Explain briefly the presence of seasonal surface streams in an area composed of chalk. (See p. 53.)

  (e) Give two reasons why chalk does not develop the same landscape features as Carboniferous Limestone. [CAM]

The work of rivers

The features made by rivers tend to be one of the most frequently examined topics among landforms. Fig. 36 explains some of the terms used in describing rivers.

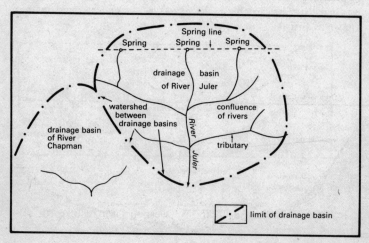

Fig. 36

Long profile

If you (i) straightened a river out, keeping the heights where the contours crossed the river, and then (ii) drew a section *along* the course of the river ... you would have a **long profile**, which might look something like fig. 37.

Fig. 37

Cross-profile

A section across a river at any point may be called a **cross-profile**.

You should realize that rivers are the most important factors in cutting a valley, but that other processes (see fig. 38) are responsible for widening the valley. The river then removes the **waste** (rock fragments and soil). These processes include:

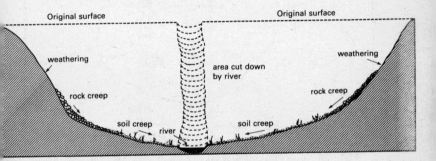

Fig. 38

weathering – breaks up rocks, creating waste:
    mechanically (i) by **frost shattering**, freezing and expanding
                (ii) by **temperature change**, hot days and cold nights
    chemically when some rocks rot under the action of mildly acid rain-water (see p. 53)
creep of rock fragments and soil brings the waste to the river
rainwash brings more material down
rills (tiny streams) carry waste, too.
The amount of water, and the rate it is flowing at, is called the river's **discharge**, and is measured in cubic metres per second or **cumecs**.

Q. 6 ... study the distribution graphs of the discharge of rivers A and B [fig. 39].

(a) *Compare the discharge of the two rivers.*

(b) *Suggest possible reasons why the discharges are so different.*   [OXF]

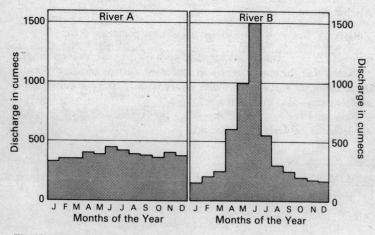

Fig. 39

Writing about distributions throughout a year

1. Look for the *maximum* (greatest) figure, name it and give the month.
2. Look for the *minimum* (smallest) figure, name it and give the month.
3. Calculate the *range* (difference between maximum and minimum).

If you are asked to contrast two yearly distributions, deal with each of these points in turn, discussing the two distributions alternately, using words like 'but' and 'whereas'.

Revision Exercise 4

1. With reference to fig. 39, fill in the gaps in the following and choose the correct alternatives, thereby answering Question 6(a):

River A has a maximum discharge of _____ cumecs in _____ (give month). B's maximum is also in _____ but is _____ cumecs, nearly _____ times as large. A's minimum discharge of _____ cumecs occurs in _____. B's minimum is _____ cumecs; $\dfrac{\text{lower}}{\text{higher}}$ than A's, but also in _____.

River A has a $\dfrac{\text{greater}}{\text{smaller}}$ range of discharges than River B – a range of _____ cumecs as compared with B's _____ cumecs range.

(Notice that the comparison also contained a rough measure of how many times one feature was bigger than another. This will gain marks in most comparison questions.)

2. Now read the box below and select the factors you think may be relevant in answering Question 6(b).

SOME FACTORS AFFECTING RIVER DISCHARGE

1. **Rainfall distribution**: can be (i) evenly distributed (the same all year round), or
   (ii) *seasonal*, e.g. more in winter
   summer monsoon
   June/July
   summer drought
   winter drought
2. **Temperatures**: can be (i) high, and evaporate river water
   (ii) low, and freeze up the water supplies; when snow-melt occurs in spring (April/May) the discharge often increases dramatically

---

Most examination questions on rivers are concerned with the work done by rivers in **eroding** (removing and transporting) and **depositing** material, and the features which result.

Erosion

1. How it is done

(i) **Solution:** dissolving material in river water.

(ii) **Hydraulic action**, as the force of the water pushes and pulls out loose material.

(iii) **Corrosion**, as the material the river has picked up collides with the bed and banks, removing more material. The material carried by a river is called its **load**.

2. What happens to the load?

(i) It is carried along by

(1) **saltation** as the larger particles bounce and jump along the stream bed

(2) **in solution**

(3) **in suspension** – most of the load 'hangs' in the water.

(ii) It is made smaller and smoother by **attrition** – as particles rub against each other.

3. How much can the stream carry?

This depends on:

(i) The volume of water (discharge: see box on p. 59).

(ii) The speed; the faster a stream flows, the more it can erode and carry. A raging flood stream can move boulders by saltation.

(iii) The cross-section. A wide, shallow stream does not flow as quickly as a narrow, deep one (fig. 40). This is because there is less friction slowing stream Y down and *faster streams carry more.*

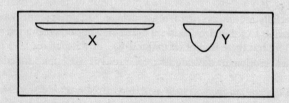

Fig. 40 *Two different cross-sections, but the same area of water*

You should be able to recognize the erosional features of a river from a photograph (see fig. 46) and on a map (figs 41 and 43). Erosional features of a river include: V-shaped valley, waterfall, interlocking spurs. There are spurs each side of the river at 852932 and 854935 (fig. 41) and they are said to interlock.

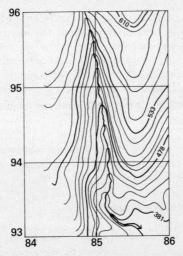

Fig. 41 *Erosional features of a river (extract from O.S. 1:50,000, Wensleydale sheet)*

Revision Exercise 5

Draw a cross-section from 845945 to 855945 (fig. 41). This will show you a V-shaped valley. (Always write the letter V with straight sides when writing about rivers. You will then avoid confusion with U-shaped valleys, p. 76.)

Waterfalls are always written as such on O.S. maps, in blue. They may be due to a band of resistant rock, slowing down the river's rate of erosion, but there is another major cause, as we shall see later (p. 64).

Vertical erosion, or downcutting, by a river is sometimes called **dissection**.

Q. 7 In a dissected plateau such as Exmoor, the rate of dissection would increase if

A. continued erosion uncovered more resistant rock

   B. *increased deposition occurred in the valleys*
   C. *sea level rose*
   D. *the area experienced a greater annual rainfall*
   E. *sea level remained unchanged.*   [LON B]
Refer to section 3, p. 60, to answer this one.

Deposition and erosion together

When a river bends, or **meanders**, there will be erosion on the faster, outside curve, forming a steep **bluff**, and deposition on the inside curve where the water moves more slowly: the **slip-off slope** (fig. 42). The reasons for the formation of meanders are not fully known. Sometimes a river 'short-circuits' a meander to leave behind an ox-bow lake (fig. 46).

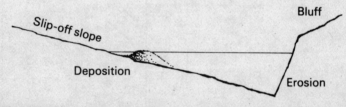

Fig. 42

Deposition

This occurs when
   (i) the volume falls – perhaps because of drought;
   (ii) the speed of the river drops – perhaps as it enters a lake or as the gradient slackens;
   (iii) the river becomes shallower and wider. See p. 60.

Depositional features
1. **Deltas** form at the entrance to lakes and seas, unless tidal currents in seas carry the sediment away.

2. **Flood plains** are areas each side of the river where river deposits – **alluvium** – have been laid down during time of flood. These depositional features may be subsequently eroded (p. 64).
3. **Levées** – where the river has actually raised its bed by deposition.
4. **Braiding** – where so much deposition has taken place that the river has had to split its course to get round the deposited material.

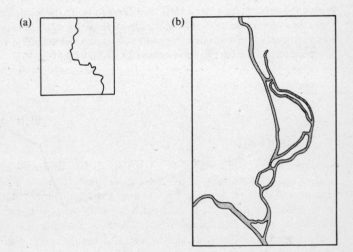

Fig. 43 (*a*) *Meanders on Holton Brook, Oxon.; (b) Meanders and braiding on the River Cherwell, Oxford*

You can see (fig. 43) that meanders vary greatly in size!

River capture

Sometimes a river which is cutting down very rapidly cuts **backwards** from its source until it links with a river which is at a higher level. The more vigorous river (X) diverts the higher-level water (Y) into its own course (fig. 44). This is known as **river capture**; it is very rarely examined and you should consult a specialist physical geography book to understand it fully.

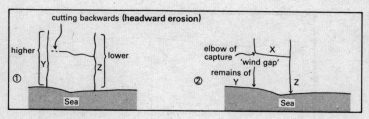

Fig. 44 *River capture*

Rejuvenation

Normally a river will continue to erode until its **long profile** resembles that of fig. 37. It is then said to have a **graded profile**. Should the land rise, or the sea fall, the river will start to cut a new long profile, starting at the mouth. It will erode into its old bed, producing **river terraces** as it cuts into its old flood plain. Where the newly eroded section meets the old, graded profile, there will be a sudden change in the gradient, known as a **break of slope**. This may result in a waterfall (see p. 61 and fig. 45). The river will also cut deeply into its meanders, forming **incised meanders**. When new power is given to a river in this way, **rejuvenation** is said to have occurred.

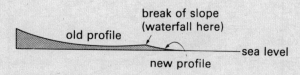

Fig. 45

FEATURES OF REJUVENATION

River terraces
Break of slope – waterfalls
Incised meanders

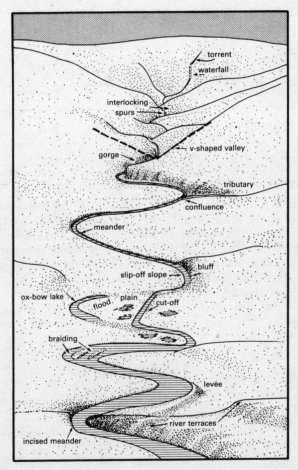

Fig. 46 *Features of the river valley*

Question 8 illustrates that two answers may be correct in a multiple-choice question. (The examiners will tell you how to indicate this.)

Q. 8 The break of slope in the long profile [fig. 47] may be due to
   A. a rise in sea level
   B. a band of resistant rock
   C. rejuvenation

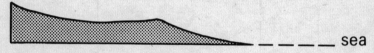

Fig. 47

Q. 9 Study [fig. 48]. One river flows over soft rock; one partly over hard.
  (a) Give technical terms for the junction of two rivers; the area drained by the river and its tributaries; the dashed line between the rivers 1 and 2
  (b) Give two differences likely to be found on river 2 at A and B
  (c) Describe and explain the feature which might occur at C
  (d) Describe what is likely to happen at D in the future (See fig. 44.)

This is a useful revision question.

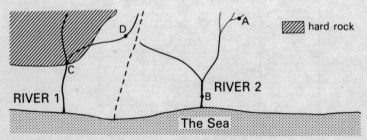

Fig. 48

Part d refers to the process of **river capture** – see p. 63.

Drainage patterns

The patterns made by rivers in an area are called **drainage patterns**. Two you should know are **radial drainage** (fig. 49), as seen in the Lake District (look at this in an atlas or, better, on O.S. sheet no. 90) and **trellised drainage** (fig. 50), found where soft and hard rocks alternate.

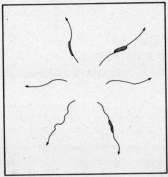

Fig. 49 *Rivers forming radial drainage*

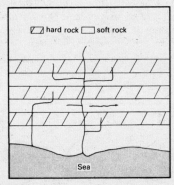

Fig. 50 *Trellised drainage*

Coastlines

Many factors have affected the development of our coastline. On the largest scale, the major factor has been a change in the level of the sea.

Changes in sea level

Sea level may
- (i) fall (1) because the land rises due to earth movements
           (2) because sea water is locked up in glaciers
- (ii) rise (1) as the glacier ice melts
            (2) because the land sinks.

Coastlines affected by rises in sea level are called **coastlines of submergence**.

Coastlines of submergence

Two main types are distinguished:

1. **Atlantic coasts** where mountains are at right angles to the sea (figs. 51 and 52).

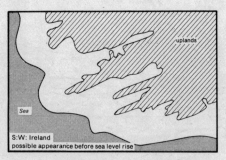

Fig. 51 *S.W. Ireland – possible appearance before sea level rose*

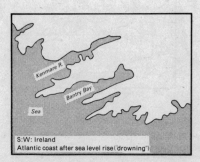

Fig. 52 *S.W. Ireland – Atlantic coast after sea level rise ('drowning')*

2. **Dalmatian (Pacific) coasts.** These occur when the mountain ranges are parallel with the coast. There is no good example in Britain, but the Californian coast is a typical case.

Smaller-scale features of drowned coasts

When a V-shaped river valley is submerged (**drowned**) it forms a long, shallow, winding inlet, called a **ria**. (Notice how often the 15 m contour runs parallel with the coast in fig. 53.)

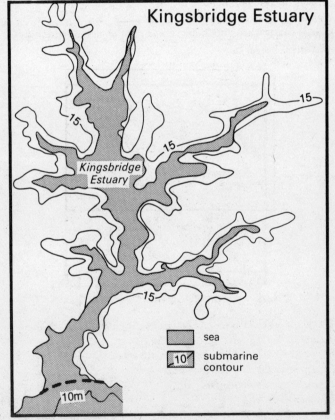

Fig. 53 *Kingsbridge Estuary, Devon – a ria (extract from O.S. 1:50,000, sheet 202)*

Revision Exercise 6

Describe Kingsbridge Estuary (fig. 53) in terms of:

   (a) distance from the estuary mouth (marked with a dashed line on fig. 53)

   (b) average width (use the R.F. for (a) and (b))

   (c) the depth of the water (look at the submarine contour)

   (d) the shape.

When a **U-shaped (glaciated)** valley is drowned, it forms a **fjord** (sometimes spelt **fiord**). These are best developed in Norway, but they also occur on the west coast of Scotland (fig. 54). Fjords differ from rias in that they have

   (i) much steeper sides

   (ii) (usually) a shallow area at their mouth.

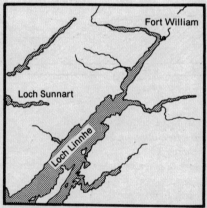

Fig. 54 *Part of the fjord coastline of the western Scottish highlands*

Q. 10 Which of the following is the correct definition of a ria:

   A. an isolated low hill

   B. the end of a peninsula cut off by wave action

   C. a river valley deepened by glaciation

   D. the drowned mouth of a river

   E. an undercut cliff.   [LON B]

Only C and D mention valleys (D by implication), and rias are not glaciated, so the answer is D.

Coastlines of emergence

Sea level may also fall, leaving the old shoreline above the present coast-line. The old shoreline is called a **raised beach** and the 'stranded' cliffs are said to be **dead cliffs**.

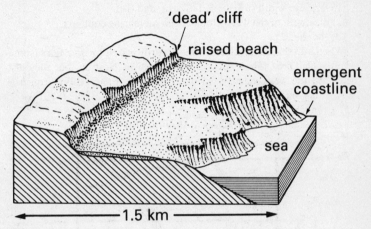

Fig. 55

Fig. 55 is taken from a London A question.

Q. 11 Describe the formation and characteristics of the landscape features depicted in the diagram.

(The labelling was *not* given by the examiners.) Remember that you must also explain the processes at work which formed the cliff before it 'died', and the beach before it was raised. For that, you must read on. To explain why sea level fell, refer to p. 68 (section i(1)).

Marine erosion

How does it take place?

The processes are very similar to those of river erosion:

    (i) **chemical action** (solution) as the sea water reacts with rocks

    (ii) **hydraulic action** as the force of the breaking waves and the **pressure of trapped air** attack weaker areas of the cliff face

    (iii) **corrosion** (see p. 60).

What happens to the material?

It is ground down by **attrition** (see p. 60) and then deposited, either on the **wave-built terrace** (fig. 56) or as part of **longshore drift** (see p. 73).

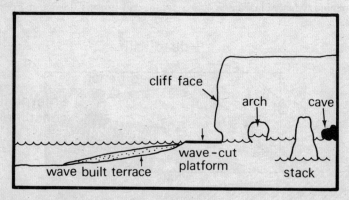

Fig. 56 *Landforms of marine erosion*

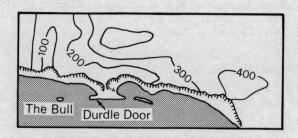

Fig. 57 *Features of marine erosion on the Dorset coast (extract from O.S. 1:25,000, Purbeck sheet)*

Cliffs

In fig. 57 The Bull rocks are stacks, and Durdle Door is a famous arch. Notice how cliffs are shown by

    (i) rock drawings

    (ii) the way the contours hit the coastline at right angles.

As a rule, resistant rocks like granite, chalk and limestone are those which form headlands (**promontories**), cliffs, arches and stacks. Remember that cliffs are also subject to the same processes of weathering and erosion which affect the sides of river valleys (fig. 38). Question 11 reminds you that marine erosion is not the only process at work on a cliff face:

Fig. 58

Q. 12 Identify the processes at work at X and Y [on fig. 58].   [LON B]
Some examining boards may ask you to describe coastal features from a map. (See Chapter 1, p. 25.)

Marine (sea) deposition

Material eroded from cliffs is ground down by attrition and then moved slowly along the coast by a process known as **longshore drift**.

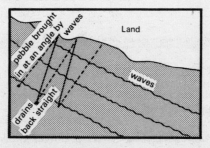

Fig. 59 *The mechanism of longshore drift*

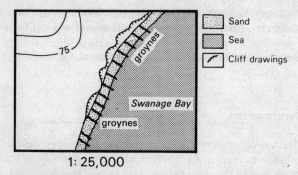

░░░	Sand
▓▓▓	Sea
⌒	Cliff drawings

1: 25,000

Fig. 60 *Groynes in Swanage Bay, Dorset (extract from O.S. 1:25,000, Purbeck sheet)*

Notice that the waves in fig. 59 are coming inshore **obliquely** (at an angle). **Groynes** (breakwaters) are built to stop sand moving along the coast – longshore drift. If the sand is not checked, it may build **spits** which can divert rivers and even link islands to the mainland (**tombolos**). Sand also tends to accumulate in bays, because the force of the waves is less there.

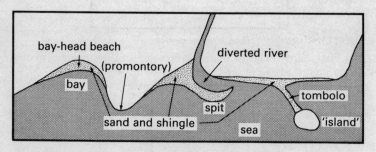

Fig. 61 *Some features of marine deposition*

Remember that erosion and deposition are linked (see p. 72 and fig. 56), and that both may work to straighten the coast by eroding promontories and filling up bays.

Q. 13 Give an illustrated account of the work of the sea in straightening an indented coastline.

[LON A]

This is a useful revision question. 'Illustrated' means using sketch maps and diagrams. 'Indented' means going in (bays) and out (headlands).

The landscape of Britain has also been affected by events in the past, notably glaciation.

The work of ice

The Ice Age occurred in Britain because the climate became colder, and snow began to accumulate. Over many years the weight of succeeding snowfalls turned the snow to ice, and, under the influence of gravity, the ice began to flow downslope to merge and form a **valley glacier**. The hollows in which the ice accumulated are called **corries** (figs. 62 and 63).

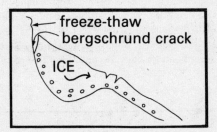

Fig. 62 *Cross-section of a corrie during glaciation*

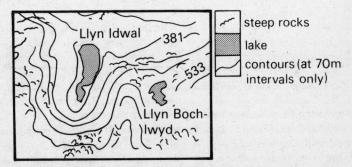

Fig. 63 *Corries in North Wales (extract from O.S. 1:50,000 sheet 115)*

Corries deepen because frost-shattered material falls down the **berg-schrund crack**. This is incorporated into the snow, which, under pressure, is turning into ice, known as **nevé**. As they move, the rock fragments and ice **scour** away the land surface in a process known as **abrasion**. Sometimes the ice freezes on to rocks (rather as your fingers stick to the ice-tray in the fridge) and pulls fragments from the rocks as it moves downslope in response to gravity. This is called **plucking**. The ridge of high land between two corries is called an **arête**. A peak surrounded by corries is a **pyramidal peak** (see figs. 64 and 65).

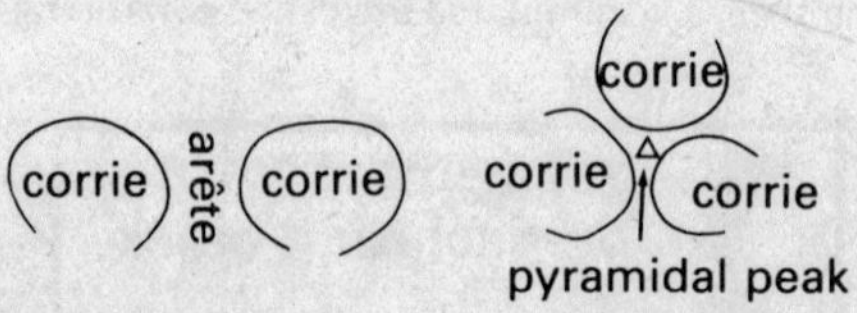

Fig. 64

Fig. 64 shows the kind of simple sketch you might use to answer Question 14:

Q. 14 ... choose two features formed by glacial erosion. Account for the formation and characteristics of the landforms you choose.

See above to 'account for the formation' of a corrie and use fig. 63 to give the *dimensions* and *shape* of the corrie which contains Llyn Idwal. Dimensions and shape are both included in the word 'characteristics'. Question 14 was based on a photograph recognition test. Fig. 65 attempts to show you the appearance of glacial features, but there is no substitute for looking at as many photographs of glacial scenery as you can. (See 'Further Reading' on p. 80.)

Valley glaciers form **U-shaped valleys**, **truncated** (or cut off) **spurs**, leave tributary valleys as **hanging valleys** high above the main valley floor, and dump **moraines** (rock and clay particles eroded by ice) along the valley side (**lateral moraine**) and at the end of the glacier (**terminal moraine**).

A **glacier** is composed of ice of varying thickness. The thicker the ice, the more powerful the erosion. As the ice moves, cracks – known as **crevasses** – open and shut, and moraine is carried along.

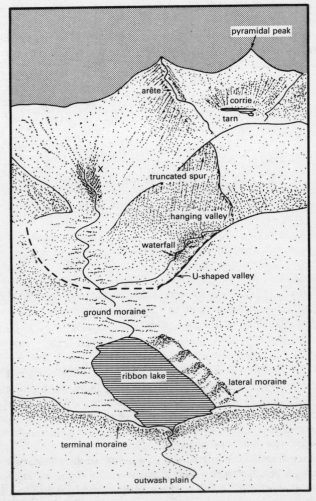

Fig. 65 *Glaciated scenery*

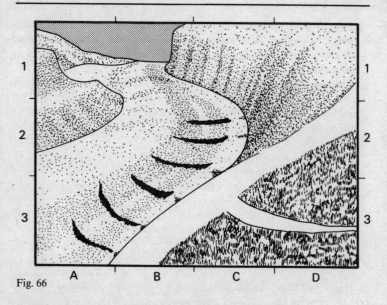

Fig. 66

Fig. 66 is based on a photograph used in an O-level question:

Q. 15 Describe carefully the main features shown on the surface of the glacier. Show how these formed.   [LON]

Do note the words 'on the surface' of the glacier. This can only refer to crevasses, shown as black cracks in the drawing, and moraines, shown by shading. Don't make the all-too-common mistake of seeing a glacier and immediately writing about corries and U-shaped valleys, even when they have not been asked for. If reference numbers and figures are given (as in fig. 66) use them in the way that you would use grid references.

For the process whereby moraines are formed and moved, re-read p. 76; to find out what a moraine consists of, read on.

Glacial deposition

This can be divided into two types:

1. Deposition which has taken place *under the actual ice mass*. This is **glacial deposition** and the mass of material deposited is known as **moraine**, or **till**, or **boulder clay**. It consists of clay embedded with sharp rock fragments of all sizes and is sometimes moulded into low, oval hills, called **drumlins**.

2. Deposition by **meltwater streams** flowing from the glacier as the climate becomes warmer. This material is found *beyond* the furthest extent of the glacier. It chiefly consists of sands and gravels and forms an **outwash plain** (see fig. 65).

The contrast between the styles of two different examining boards can be seen in the extremely detailed knowledge of processes you need for Question 16, and the way the examiner tries to link landforms with man's response to them in Question 17.

Q. 16 (a) *What is the snow line?*   [3]
   (b) *Why does its position vary?*   [3]
   (c) *Describe how a valley glacier may be formed.*   [3]
   (d) *Contrast plucking and abrasion.*   [3]
   (e) *When does a glacier begin to retreat?*   [3]
   (f) *Explain why the floors of glacial valleys erode unevenly.* [3]
   (g) *Describe the nature of deposits which form a terminal moraine.*   [3]
   (h) *Explain how a terminal moraine may be formed.*   [3]
                                     [CAM]

Question 17 begins with a photograph much like fig. 65:

Q. 17 (a) *With the aid of an annotated diagram summarize the evidence which suggests the landscape is largely the result of ice action.*   [5]

For this, the diagram alone, suitably annotated, could be enough. 'Annotated' simply means 'labelled'.

Q. 17 (b) *(looking at a town in the centre of the photograph)*
   (i) *Suggest reasons for the town's location.*
   (ii) *What are the probable functions of the town?*   [4]
  (c) *The land-use on opposite sides of the valley is different.*
   (i) *Describe the contrasts in this land-use on the lower valley sides.*
   (ii) *Suggest an explanation for this contrast.*   [3]
                                     [OXF]

For Question 16, you need a more detailed book than this. For Question 17, see Chapter 4 of this book.

Do make sure you know the type of landform questions your examination board is likely to set.

Questions are also asked on landforms not seen in Britain, and these landforms are dealt with in Chapter 7, under 'World Relief'.

Rivers, coasts and ice are the 'big three' in landforms. If you know these *really well* you should cope with landform questions satisfactorily.

Further Reading

Principles of Physical Geology by Arthur Holmes (3rd ed., Nelson, 1978) is an A-level book. If you can find a library copy, *don't read it*, just use it for the excellent photographs in Chapters 18, 20 and 23.

4. British Isles: Human Geography

British Isles: Population

The average density of population in Britain is about 318 per square kilometre. Fig. 67 shows the distribution of population, which reflects
negative factors, e.g. mountainous relief
                         bleak climate
                         poor soils
                         lack of resources
positive factors, e.g. gentle relief
                         warm, moist climate
                         mineral resources
                         industrial development.

Sparsely populated areas

The Highlands of Scotland, for example, have been losing population for centuries. Negative factors are:
   (i) mountainous relief: most of the land is over 400 m
   (ii) bleak climate: rainfall everywhere between 750 mm and 2,000 mm; altitude lowers temperatures (p. 41); there is on average less than an hour of sunshine daily
   (iii) the best soils were removed by glacial erosion.
Traditionally, the inhabitants lived on **crofts**: farms with a tiny arable acreage, growing potatoes, hay, turnips and oats. Cattle and sheep are pastured on hill land. Crofters need an additional income: a road-mending job, weaving, etc.

The poverty of this way of life has forced many crofters to move. Most crofters' children must go to cities for their schooling; they rarely wish to return. They are attracted to the towns by
   (i) industrial jobs, e.g. oil-related industry at Aberdeen
   (ii) tourist employment in resorts such as Oban or on Clydeside
   (iii) better **services** – educational, medical, power supplies, shops.

Other upland areas are losing population; questions asking for their locations and the reasons for population loss are especially common on Oxford board papers.

Revision Exercise 1

Identify sparsely populated regions 1–10 (fig. 67). What physical features have they in common? Do not be vague: give figures.

Other areas losing population include declining industrial regions, like the valleys of south Wales, and inner city centres.

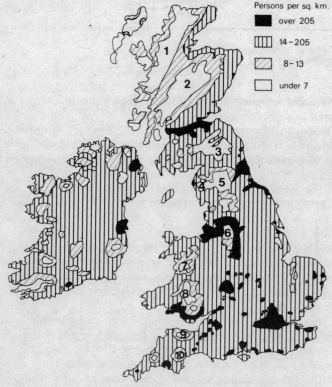

Persons per sq. km.

over 205

14 – 205

8 – 13

under 7

Fig. 67 *British Isles – population density*

Rural settlement

Population densities are low in rural areas: around 60/km² in Britain.

Villages are often located along a physical feature such as a spring line (fig. 68). Rural settlement is often described by the pattern it makes:

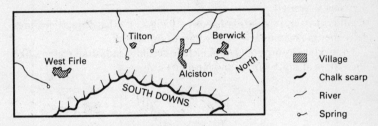

Fig. 68 *Part of the scarp of the South Downs, Sussex*

Nucleated settlement has most of the people living in a village with some scattered farms.

Dispersed settlements have no central village but are spread over a large area. These patterns often result from particular social systems, in many cases centuries old (fig. 69).

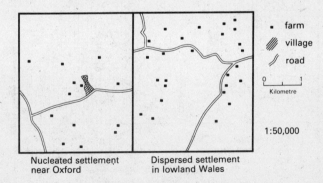

Fig. 69 (*a*) *Nucleated settlement near Oxford;* (*b*) *Dispersed settlement in lowland Wales*

Many villages are losing inhabitants, who are attracted to towns by the factors given on p. 81. Other villages grow, as richer city workers look for a house in pleasant surroundings and **commute** to work daily.

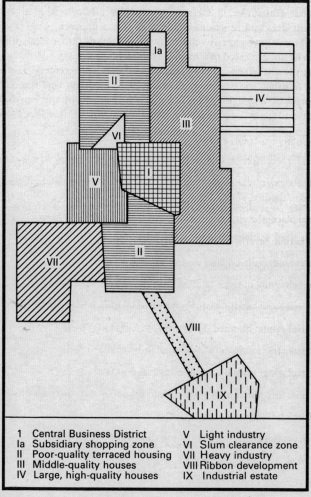

1 Central Business District	V Light industry
Ia Subsidiary shopping zone	VI Slum clearance zone
II Poor-quality terraced housing	VII Heavy industry
III Middle-quality houses	VIII Ribbon development
IV Large, high-quality houses	IX Industrial estate

Fig. 70

Urban settlement

The **functions** of a town are the purposes it serves and may include trade, welfare, administration, education, industry, finance, retailing and housing.

Walter Christaller, a German geographer, thought that towns of different sizes and functions would arise, i.e. a network of a few cities, several large towns, many smaller towns and so on. Other ideas of **urban hierarchy** are discussed in Chapter 6 (under 'City sizes').

Many towns **serve** an area and may be called '**regional centres**'. Different urban **functions** may result in different areas of **urban land-use**, known as **functional zones**. Fig. 70 opposite is a diagram of the **morphology** (shape) and functional zones of a regional centre in the developed world.

Vocabulary for town studies

By-pass: road built on one side of a town so traffic can avoid the centre

C.B.D.: *Central Business District* – contains government and business offices, finance, shops, theatres

Central place: the city, because it is central to the area it serves

Conurbation: several cities, merging into a continuous built-up area

Derelict land: built-up land no longer used

Green belt: zone of open land around the city

Heavy industry: smelting of metals; chemicals and oil refining

Industrial estate: planned development of many factories

Light industry: manufacture of small industrial products

Manufacturing: production of industrial goods

Pollution: introduction of poisons into air and water, often as waste

Residential: housing

Ribbon development: building along the line of a main road

Ring road: by-pass encircling a town

Slum: very poor housing

Slum clearance: demolition of slums

Subsidiary: of secondary importance

Suburb: area of more spacious housing near city boundary

Terraced houses: a row of houses joined together

Trading estate: as 'Industrial estate'

Urban renewal: improving inner city area by slum clearance and rebuilding

Zone: section of a town having a distinct function

Q. 1 (a) (i) Define 'regional centre'.

(ii) State, giving a reason in each case, which of the zones shown in the diagram is likely to have

(1) the highest density of resident population

(2) the lowest density of resident population

(3) the greatest concentration of shops and offices

(4) the greatest amount of derelict land, and

(5) the greatest problem of industrial pollution.

(b) Using a sketch map to illustrate your answer, describe and explain the functional zones of any one named capital city in the developed world. [LON]

Generally, the better the housing the lower the density of population. The section on vocabulary should help you with the rest.

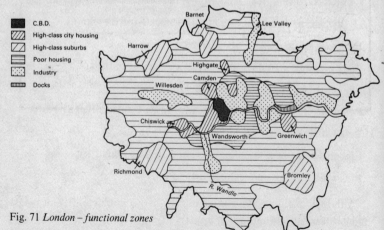

Fig. 71 *London – functional zones*

Revision Exercise 2

Fig. 71 is not a sketch map. Make one from it by drawing

(a) the River Thames

(b) the C.B.D.

(c) bands of industry (i) each side of the Thames east of the C.B.D.

                      (ii) along the Lee valley

                      (iii) along the Wandle valley

(d) 'high-class' suburbs

(e) poor housing each side of industrial zone

(f) London's only remaining dock area: the Albert and George V docks.

You can explain that high-class residential areas are scattered throughout the city, generally on the *higher land*.

Conurbations

Fig. 72 shows the West Midlands industrial conurbation.

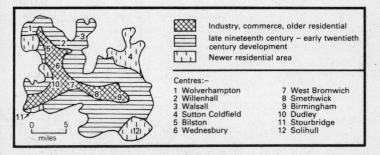

Fig. 72 *The West Midlands industrial conurbation*

Although one town runs into another, there are many different types of land-use: some are shown in fig. 71.

Every conurbation has **different reasons for growth**.
For the West Midlands these are:
  (i) historical – centre of canal network
               – South Staffs coalfield
               – iron and steel industry
               – development of skilled labour
  (ii) skilled labour used in engineering, car manufacturing.

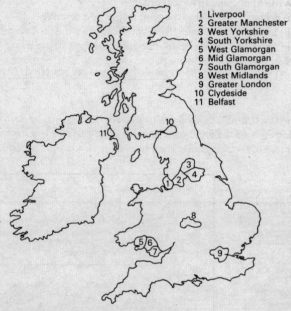

1 Liverpool
2 Greater Manchester
3 West Yorkshire
4 South Yorkshire
5 West Glamorgan
6 Mid Glamorgan
7 South Glamorgan
8 West Midlands
9 Greater London
10 Clydeside
11 Belfast

Fig. 73 *Conurbations in Britain*

Every conurbation has **the same problems**:
(i) derelict land from declining industries
(ii) Victorian slums
(iii) traffic congestion
(iv) industrial pollution
(v) lack of open spaces.

Conurbations are losing population (i) as some industries decline; (ii) as people move out to more pleasant suburbs.

Q. 2 By means of named examples, write briefly on
  (*a*) *urban hierarchies*
  (*b*) *functional zones of towns*
  (*c*) *conurbations ...*

[LON A]

'WRITE BRIEFLY'

Write briefly and in logical order:
(i) **define** – say what the term means and **describe**
(ii) **locate** – give an example, preferably with sketch map
(iii) **explain** – the processes that have gone on, or are going on.
Note any problems in a separate paragraph.

Towns

During this century, many **dormitory towns** have grown up within about 50 kilometres of the large conurbations. These are towns where a large percentage of the population commutes daily to work in the conurbation.

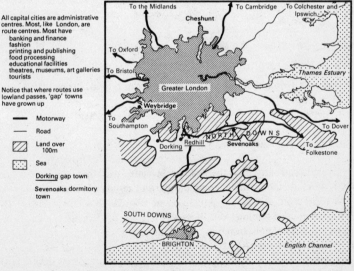

Fig. 74 *Selected features of development around London*

Q. 2 (d) [Explain what you understand by] gap towns.
Use the instructions in the 'Write Briefly' box (p. 89) and the information in fig. 74. The sketch map should show the gap town in relation
to London, the route, the destination, and the high land.

Rivers also influence town growth. Crossing points are important,
especially the **lowest bridging point**: the crossing nearest to the open sea.
If ships come up the river from the sea, the town at the **head of navigation**
of the river, i.e. the highest point which can be reached by cargo vessels,
will be an important port. London fills both these roles for the Thames.

New Towns
Britain is short of housing, but to cut down pollution and provide open
space for leisure, planners like to keep a belt of undeveloped land around
cities: a '**green belt**' (fig. 75).

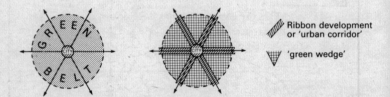

Fig. 75 *Green belts and green wedges*

If people living in slums are to be rehoused, but the green belt is not
to be built on, there are two possibilities:
   (i) **urban renewal** (p. 134), also known as **redevelopment**
   (ii) **New Town development**.
New Towns are planned – away from built-up areas
                    – with more living space and fresh air
                    – with factories, offices, shops and schools to
                       provide jobs and services.
A map with explanatory labels on it, like the one shown in fig. 76
is known as an **annotated sketch map**; it is almost sufficient as an answer
to part (c) of the next question, as well as to part (b).
*Q. 3 (a) Describe the distribution of New Towns and give reasons for their
creation.*
  (b) Describe the layout of one named New Town.

(c) *What are the attractions of New Towns, for people and for new industrial developments?* [CAM]

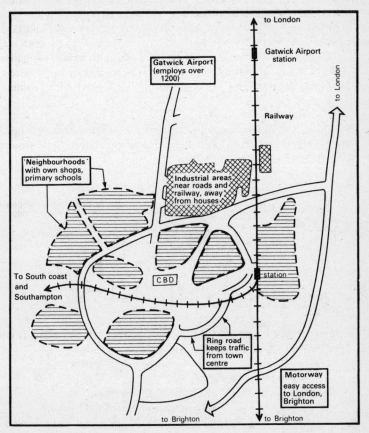

Fig. 76 *Crawley, a New Town*

On the sketch map you should clearly distinguish between the attractions for (i) people and (ii) developing industries, perhaps by using two different colours for the labelling, and indicating this in a key or in your written account.

Crawley's industries include engineering, plastics, printing, electronics, food and drugs (light industries). Notice that the industrial area is separate from housing, unlike the older industrial towns. Every effort is made to keep heavy traffic from housing areas, and New Towns have much more open space than old ones. Fig. 77 shows some of Britain's New Towns.

Fig. 77 *New and expanded towns*

To some extent New Towns have failed: many are near enough to the conurbations for people to use them as dormitory towns and they are expensive to build. Recent government policy has been to **expand** existing towns like Telford or Milton Keynes – made from a group of villages. This is cheaper than creating an entirely new town.

British Isles: Mining

Coal

Coal formed in **Carboniferous** times, and the **seams**, varying in thickness from a few centimetres to several metres, are interlayered with other sedimentary rocks, like clay and sandstone. Most coal seams have been folded and eroded (fig. 78).

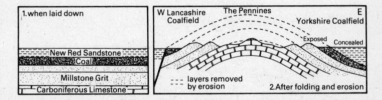

Fig. 78 *Coalfields on the flanks of the Pennines*

The coalfields to the east of the Pennines have thick, unbroken seams which dip underground eastwards, and are mined at a depth of over 2,000 ft on the **concealed** coalfield. These deep collieries are larger than those on the **exposed field**. To the west of the Pennines, the Lancashire coalfield has no concealed field: the seams have been 'cut off' by faults.

The South Wales coalfield is much more difficult to work because the thin seams have been folded and faulted (fig. 79). Here there are no broad seams, easily worked, but narrow, **contorted** seams, which stop and start abruptly. Machinery cannot be profitably used, as it can to the east of the Pennines.

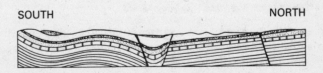

Fig. 79 *North/south section through the South Wales coalfield*

Q. 4 (*a*) *Draw a diagram to show why there are coalfields on both flanks of the Pennines*
(*b*) *Draw a diagram to show how features of coal seams may make mining difficult.* [Half question]

NOTE. No writing is asked for but you can annotate your diagrams.

Coalfield developments

Nineteenth-century coalfields developed **heavy industry**:
    iron and steel smelting
    heavy engineering
    shipbuilding (coastal coalfields)
    textiles or chemicals.

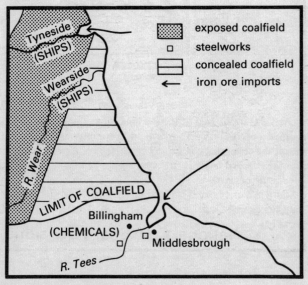

Fig. 80 *The Northumberland and Durham coalfield*

With it came spoil heaps and smoky factories. Newer coalfields are more mechanized and have polluted the environment far less.

Fig. 81 shows the main areas underlain by coal in Great Britain. You may be asked to draw this, but there should be an outline map to help you.

Fig. 81 *National Coal Board Divisions*

Coalfield problems

All over Europe old coalfield areas are in decline because

1. Demand for coal has fallen: oil and natural gas are increasingly used to make electricity, run ships and trains, heat homes.
2. The easiest coal seams have been worked out: mining is more difficult and more costly.

3. Machines are replacing men.

In addition the heavy industries – notably iron and steel, and shipbuilding – are in decline because of foreign **competition** and the present **recession** in world trade.

Other important mineral resources

Iron ore
Salt – for the chemical industry
Gypsum – for plasterboard, chemicals
Kaolin – the raw material for pottery
Limestone – for cement-making.
You need to know the locations of all these to answer Question 5.

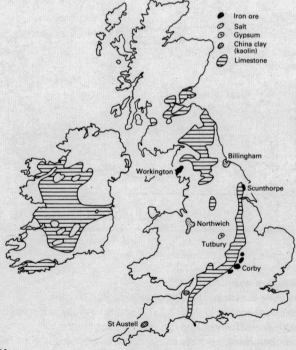

Fig. 82

Q. 5 Which of the following sets of mineral deposits is found at the three locations marked on the map of England and Wales [fig. 83]?

   A. China clay; coal; tin

   B. Iron ore; limestone; china clay

   C. Iron ore; salt; china clay

   D. Iron ore; coal; salt

   E. Coal; salt; tin   [LONB]

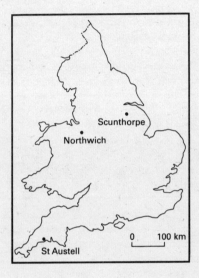

Fig. 83

British Isles: Power

Electricity

Although some coalfields are in decline (pp. 95–6), coal is still of vital importance: it is burned to produce one third of Britain's electricity in **coal-fired thermal (or steam) power stations**. Since coal is bulky, and usually moved by rail or canal, most coal-fired power stations are

   (i) near coalfields

   (ii) near navigable waterways

   (iii) near vital cooling water

   (iv) near large towns.

The valley of the Trent fulfils these factors and is sometimes known as 'the river of power'.

Oil-fired thermal power stations are frequently located on the coast since much of the oil used is imported from the Middle East. Power stations often appear on 'photograph' questions.

Pump storage schemes. Electricity cannot be stored. At Ffestiniog, in North Wales, surplus night-time electricity is used to pump water up into a high-level lake. During daytime the water runs downhill to generate hydroelectric power (fig. 84). This station uses up unwanted electricity, but consumes more than it produces.

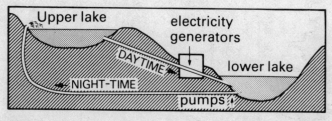

Fig. 84

Nuclear power may be used to heat the steam which turns the generators. Such stations are located
– near large supplies of water (usually on the coast)
– away from large centres of population (for safety).
You should be able to recognize the pattern of nuclear power stations on the map, and to duplicate the map, but you need not know all the names. One or two examples will suffice. Multiple-choice questions frequently give a map with dots, and ask what the map shows, e.g. ship-building yards? oil refineries?

A map of nuclear power stations will show the inland site of Traws-fynydd, North Wales: look for that.

Once built, a nuclear power station runs very cheaply; thus it is expected that Britain will generate increasing amounts of power by this method in the future. We also sell nuclear power plants.

Fig. 85 *Great Britain – nuclear power stations*

Hydroelectric power is produced when water flows to turn **turbines**, i.e. generators. Some **H.E.P.** is produced in North Wales, but the major production area is the Scottish Highlands.

An H.E.P. system: Loch Tummel, Scotland (fig. 86)
1. Suitable physical factors:
   (i) precipitation is heavy – 1,500–1,700 mm
   (ii) finger-lakes in glaciated valleys act as natural reservoirs
   NOTE. Steep slopes are *not* vital to H.E.P.

2. Human factor: Loch Tummel is near consumers in Perth and Dundee.

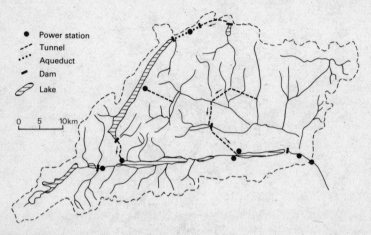

Fig. 86 *Loch Tummel H.E.P. scheme*

Revision Exercise 3

Describe the Loch Tummel scheme in terms of area; number of lakes, tunnels, aqueducts and dams, and the physical and human factors that aid its development. Remember that the whole Loch Tummel scheme only generates as much power as *one* large coal-fired station.

All electricity is fed into a system of cables, carried by pylons, called the **National Grid**.

Natural Gas

All gas used to be made from coal, but in 1967 Britain's first natural gas came ashore. (Another reason why the demand for coal has fallen.) *Q. 6 The map [fig. 87] shows the main gas pipelines in Britain.*

   (a) What are the main sources of Britain's gas supply?
Use an atlas to add these gas fields to the map: Frigg, Leman Bank, Indefatigable, Viking, West Sole, Hewitt. Learn the approximate location of the fields, and *one name* as an example. If your atlas is not up to

date enough to show these fields, go to your local library and ask the librarian for help.

Fig. 87 *North Sea gas*

Q. 6 (b) Describe and explain the chief features of this network.

[OXF, half]

Having located the gas fields, describe:

(i) the *incoming* pipelines

(ii) the main *destinations*

(iii) regions where the network is (1) most dense; (2) sparse, and explain why.

Remember, the lines go to where the demand for gas is greatest. Because gas is easily transportable it is taken to industrial areas: it has not created industrial areas in the way coal did.

Mineral oil

Mineral oil is vital to twentieth-century life. Its uses include:
(i) lubricant
(ii) fuel – for transport, power stations, heating
(iii) raw material – for plastics, textiles, chemicals (p. 109).

Oil terminals

A 200,000-ton supertanker carries oil most cheaply, but needs 16 m of clear water. Special deep-water terminals have been constructed at Loch Long, Scotland; Milford Haven, Wales; Bantry Bay, Eire.

Oil is then sent to **refineries** by smaller tankers or by underground pipeline. Oil refineries are shown in fig. 88. Again, learn the pattern they make, and *one named* example. Most oil refineries are on deep-water coastal sites, with plenty of flat land for industrial development.

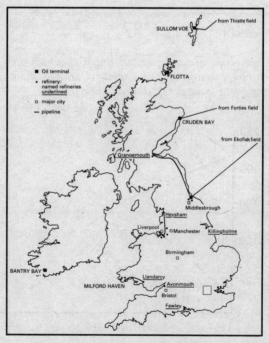

Fig. 88

Summary

Q. 7 Locate the chief sources of power in one named *country and explain their relative importance.*   [LON A]

This is best answered by making a revision map, using the text and an atlas. On an outline map of Britain:

1. Draw in the River Trent. Write the words THERMAL POWER between Nottingham and Gainsborough.
2. Add Loch Tummel (near Pitlochry) and Ffestiniog: H.E.P.
3. Add details of coalfields; natural gas; oilfields.
4. Add nuclear power stations.

The more you use your atlas in exercises like this, the better your knowledge of the British Isles will be.

NOTE. **Relative** importance means 'in comparison with each other'.

Revision Exercise 4

Using fig. 89,

1. List the four power sources in order of importance for 1975.
2. For each power source in turn, summarize how its importance has changed and look through this section to find reasons.

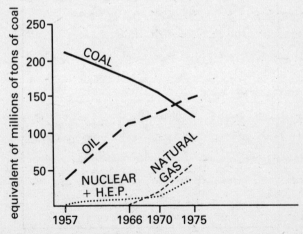

Fig. 89 *Britain's power use, 1957–75*

British Isles: Industry

Questions on this topic appear regularly in every board; possibly under
'The Developed World' section. Questions are of three types:
  (i) based on theory
  (ii) concentrating on one or two industries
  (iii) looking at the various industries in one region.
First, make sure you will understand the question by learning the
vocabulary.

Industry: Vocabulary

Assembly industry: parts made in various locations brought to one place,
  notably 'Motor-vehicle assembly'

Components: parts

Concentration: very large number of, for example, factories in an area

Consumer goods: goods bought for use in the home

Crude petroleum: oil before refining

Diversification: introducing different industries to avoid being too
  dependent on one

Electronics: small, complex, electrical goods, e.g. calculators

Environmental problem: pollution by chemicals, noise

Heavy industry: goods of large size, simple structure, e.g. girders

Inertia: industry remaining in an area long after original locational
  factors cease to operate

Integration: linking of firms engaged in same type of industry

Light industry: goods of small size, complex structure, e.g. knitwear

Locational factor: cause influencing the situation of, for example, industry

Manufacturing industry: production of goods, also **secondary** industry
  (p. 105)

Market: people wishing to buy products (**demand**)

Motor-vehicle assembly: car works

Pharmaceuticals: medicines

Recession: sharp fall in trade and employment

Reservoir of labour: pool of workers available

Smelting: extraction of metal from ore by melting

Specialization: concentration on one type of, for example, manufacturing

Synthetic: man-made

Textiles: cloth

This vocabulary is frequently tested in multiple-choice questions:

Q. 8 Stoke-on-Trent is dependent on a small range of industries and there is a danger of severe unemployment in the area if the demand for their products falls. The best solution in this case is probably the encouragement of industrial

    A. inertia

    B. diversification

    C. specialization

    D. integration

    E. recession. [LON B]

Industry used to mean manufacturing: the making of goods. Now it means all types of economic activity, e.g. 'the tourist industry'.

Primary industry is the collection of naturally occurring materials without changing them, e.g. mining (**extractive** industry), fishing.

Secondary industry is where these **raw materials** undergo change.

Tertiary industry is concerned with distribution: transport, wholesaling and retailing, and may include financing; '**more developed**' countries have more tertiary industry.

Q. 9 Lumbering, aluminium smelting, banking, electronic calculator manufacture, fish processing, zinc mining, hotel management and catering, soft drinks, natural gas

    (a) Classify each of these industries under the heading primary, secondary, tertiary

    (b) Suggest three other likely tertiary occupations in this country

    (c) Why are tertiary occupations more important in the developed world than in the developing world?

    (d) What factors affect the location and development of tertiary industry in an area of the developed world that you have studied? [LON A]

Locational factors in industry

There are four considerations:
    Power
    Raw materials
    Markets
    Labour,
    Transport to link them (the 'Five Factors'; 'PeRaMbuLaTor' is a useful memory-jogging word). The importance of each of these factors varies with different industries.

1. Power

Coal- and oil-powered industries tend to be tied to their fuel source, i.e. to coalfields or – if the oil is imported – to the coast.

Aluminium smelting needs large amounts of electricity and is located near cheap H.E.P., e.g. in Anglesey, close to the Ffestiniog station.

Changing power supplies led to relocation of factories in Britain (fig. 90).

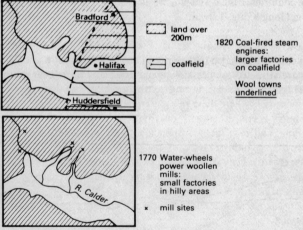

Fig. 90 *The Yorkshire woollen industry*

2. Raw materials

Industries using
    – large amounts of bulky raw materials, e.g. cement-, brick-, iron- and

steel-making, chemicals
– perishable raw materials, e.g. food canning, freezing, processing
locate near raw materials because money would be wasted by transporting
all the material which will be thrown away after processing.

Water is an important raw material, particularly in paper manufacture.
The importance of factors (1) and (2) is declining as transport improves
and factors (3) and (4) become more important.

3. Markets
A firm's market can be the public or another firm. Goods can be transported more easily these days and consumer goods are much more
important. More factories locate near the market, especially with high-
value, low-weight goods like electronics.

Markets are the major factor in tertiary industries – shops, hotels,
banking.

4. Labour
With increasing automation labour is less important than it was ten years
ago. Some countries in the **developing world** base their industries on cheap
labour, e.g. Hong Kong, Taiwan.

5. Transport
Transport costs are included in the price of any item. Market-based, light
industries are attracted to accessible areas, especially those near motorways (p. 117).

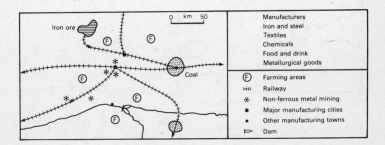

Fig. 91

Q. 10 (a) (i) From the evidence on [fig. 91], state three factors which appear to have affected the development of industries.

This part-question tests your understanding of the factors above by using a theoretical example.

HINTS

(i) Choose three different types of factors – (1), (2) and (5).

(ii) *Use the scale* – e.g. 'the two cities are equally distant (about 70 km) from the iron-ore deposits and the large coalfield'.

(iii) Don't neglect agricultural raw materials.

(ii) Briefly state what additional information you would require to explain the industrial development more fully.

HINT. Think about *people* and factors (4) and (5).

Individual industries

Five industries are summarized in the table below. Choose at least TWO and make sketch maps like the one below (fig. 92) to illustrate the locations and factors involved. Use a good atlas and figs. 71, 72, 77 and 79 where relevant.

Industry	Chief locations	Major factors*	Changes and problems
Steelmaking	Ravenscraig (Scotland) Teesside Scunthorpe Sheffield	*19th century* local iron ore; local coal	*since 1975* competition; recession (Consett, County Durham, works closed 1981 – 3,000 jobs lost)
	Port Talbot (Wales) Newport (Wales)	*20th century* port sites develop because imported iron ore is now used; skilled craftsmen and existing plant	
Ship-building	Estuaries: Clyde, Tyne, Wear	(i) closeness to steel-making areas	(i) Japanese competition
	Belfast Lough Barrow-in-Furness	(ii) tradition (iii) government aid to ease unemployment	(ii) estuaries too small for modern ships

Industry	Chief locations	Major factors*	Changes and problems
Chemicals	Tees estuary	raw materials: coal and salt (Pennine limestone); later imports	
	Merseyside: St Helens, Widnes, Runcorn	raw materials: coal water transport: Manchester ship canal for imported raw materials	
	Cheshire: Northwich, Middlewich	raw materials: salt	
Petro-chemicals	Merseyside: Ellesmere Port		Growth industry because of demand for plastics, detergents, paints, fertilizers
	Tees estuary	oil pipeline from Ekofisk	
Motor Vehicles	Bathgate (Scotland) Merseyside Birmingham Coventry Oxford Luton Dagenham	(i) easy access from other industrial areas because different components are made in different places and then assembled (ii) markets – this is a consumer industry	(i) lack of room in some locations (ii) scale – British factories too small to be efficient (iii) automation – cars 'made by robots' don't use many people; job losses. (iv) ferocious foreign competition – especially from Japanese motor-cycles

Industry	Chief locations	Major factors*	Changes and problems
Textiles	cotton: Lancashire, e.g. Nelson, Colne wool: west Yorkshire, e.g. Bradford, Halifax man-made fibres	*past factors* soft water, coal, local wool, local inventors *present factor* industrial inertia (p. 104) (i) skilled labour of Lancs. and W. Yorkshire (ii) petrochemical industry supplies fibres (iii) development areas (see p. 113)	*competition* (i) mainly from man-made fibres (ii) also from abroad

*NOTE. Don't give past factors unless specifically asked to.

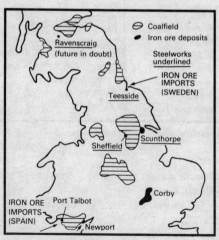

Fig. 92 *Sketch map of iron and steel in the U.K.*

You may need to know how steel is made, how salt is mined, how oil is refined ... and so on. If so, you must consult a specialist book. Ask your teacher.

Competition has been mentioned in four of these industries. This can be from other countries. Since we joined the E.E.C., we have imported increasing amounts of European goods. Japan has become a major competitor because of the efficiency of her labour force and excellent design ideas. Countries with **cheap labour** (low wages) undercut British prices.

Competition can also come from newer products. This is particularly the case in the textile industry. Man-made fibres like nylon, polyester and Acrilan are cheap and convenient – they drip-dry.

Let us see how these trends and theories work out in practice by looking at two industrial regions; first a coalfield region, typical of nineteenth-century industrial development, and second an estuary, a typical location for twentieth-century development.

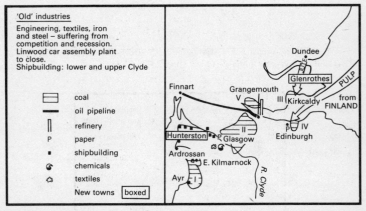

Fig. 93 *Central Scotland – industry*

Central Scotland
Q. 11 (a) Explain the factors leading to the growth of Central Scotland as a major industrial area

1. Coal: I, Ayr; II, Central; III, Fife; IV, Lothian; V, Alloa fields
2. Local iron ore – now exhausted
3. Estuaries – shipbuilding

} heavy industry – girders, bridges

4. Port sites – port industries: grain, flour, paper, soap (same pattern as Thames-side)
5. Local wool – woollen textiles, developing into cotton goods.
   NOTE. Examiners expect at least three factors for any industrial centre.

Q. 11 (b) What are the advantages and disadvantages of its continued industrial development ...?

Disadvantages: high transport costs to English markets
                 congestion
                 coalfields nearing exhaustion
Advantages: oil terminal at Finnart (Loch Long)
                 cheap H.E.P. (see p. 99)
                 refinery at Grangemouth
                 'reservoir' of skilled labour.

Q. 11 (b) cont. ... and for the establishment of new towns?   [CAM]
Disadvantages: distance from markets
Advantages: cheap land
                 demand for housing
                 demand for jobs.

New towns established include Hunterston, Glenrothes, Ravenscraig.
New industries: electronics and consumer goods because they are easily transportable; large local market.

London and the Thames Estuary
This area has the greatest concentration of manufacturing industries. This can be explained by working through the 'Five Factors' (see p. 106):
1. Power
   (i) natural gas terminal at Canvey Island
   (ii) oil refinery at Canvey Island
   (iii) eight thermal power stations between Battersea and Tilbury.
2. Raw materials
   (i) local gravel, chalk, clay
   (ii) imported via docks and riverside wharves.
3. Markets: the G.L.C. area now contains eight million people, and unemployment in the capital is relatively low; prosperity high.
4. Labour: a large number of women are available to work in light industries.
5. Transport: London has good communications by land: eight major railway termini, eight motorways; it also has a little-used canal system and two airports.

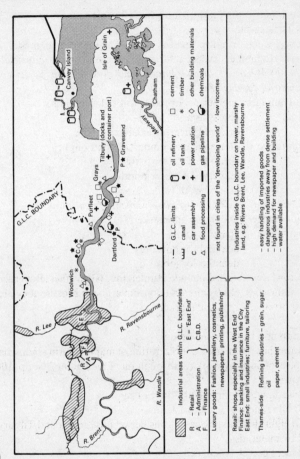

Fig. 94

Government policy

Industries are declining and unemployment is high, especially in regions north of Oxford. Governments encourage firms to move to declining areas (called **Development Areas**), by giving:

- cheap factory sites
- rate rebates
- grants towards removal costs.

This is done in order to:
- distribute jobs more evenly over Britain
- diversify industry in the Development Areas (fig. 95).

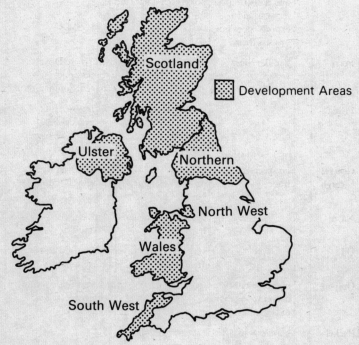

Fig. 95 *Development areas*

British Isles: Transport

Ports

Ports generally grow up
(i) where a natural relief feature – headland, bay or estuary – affords protection
(ii) where the **hinterland** is a prosperous trading area. A good port site is not enough on its own.

Type of port	Description	Locational needs	Examples
Entrepôt	port which collects and redistributes imports and exports, does not process them	good internal transport system; flat land for warehouses	London
Ferry	handles short voyages	closeness to Europe	Harwich, Dover, Folkestone, Newhaven, Southampton
Fishing		small harbour, closeness to North Sea and S.W. fishing grounds	Grimsby, Hull, Aberdeen, Peterhead, Fraserburgh, Lowestoft
General cargo	no specialized trade	closeness to prosperous hinterland	Felixstowe, Avonmouth, London
Naval		deep water, sheltered harbour	Portsmouth, Plymouth
Oil		very deep water (over 16 m)	Finnart, Milford Haven, Bantry Bay
Outport	new port built nearer to sea than old, cramped port		Avonmouth outport to Bristol
Packet	handles mail	as passenger ports, ferry ports	
Passenger	handles travellers	easy access from urban centres	Liverpool, Southampton, Tilbury
Shipbuilding		closeness to iron and steelworks, deep water	Clydeside, Teesside, Barrow-in-Furness

Q. 12 In the British Isles, name
 (a) a major oil port, and explain its advantages for this trade
 (b) a passenger port for the continent and state its importance
 (c) a port dealing in general cargo, and describe the nature of its trade
 (d) a fishing port, and describe the activities with which it is connected.

[OXF]

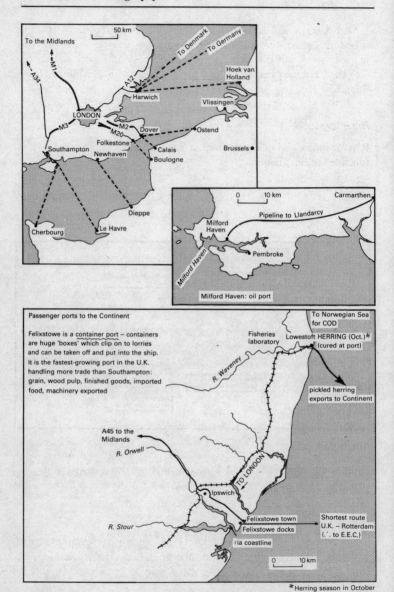

Fig. 96

*Herring season in October

Sketch maps will go a long way towards answering this – take the relevant information from the maps in fig. 96. *Use the scales* on these maps to work out distances, e.g. from the passenger port you choose – to the Continent and to London; length and width of Milford Haven Sound.

River transport

River transport is not greatly used in Britain, but fig. 97 shows the many functions of the Thames.

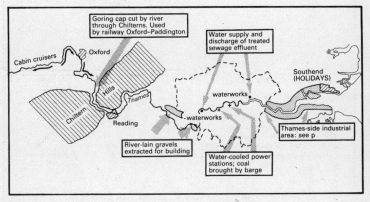

Fig. 97

Roads

The summary of British industry (pp. 111–12) indicated the importance of motorways, especially for light industry, notably electronics, and for new towns. You should know the motorway map of Britain and be able to describe the **network**.

Q. 13 The map [fig. 98] shows the network of principal roads . . .
   *(a) in what ways does the system in the south-east region (A) differ from
   that in the Midlands (B)?*

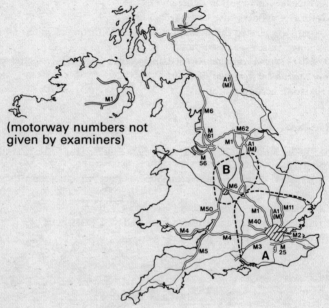

Fig. 98

This is a description question and the easiest way to describe is to *draw*
– make a sketch map (fig. 99).

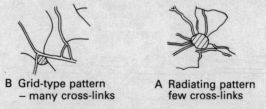

Fig. 99 *A, radiating pattern (few cross-links); B, grid-type pattern (many cross-links)*

Q. 13 (b) How is the network influenced by physical features?
Roads were greatly influenced by relief
  – avoiding uplands
  – using 'gaps'
  – focusing on lowest bridging points
Motorway builders have better technology
  – gaps are blasted through uplands e.g. M40 through Chilterns
  – long bridges span estuaries
Q. 13 (c) Name a major port within the area and discuss the adequacy of the road system that serves it [OXF]
Felixstowe needs a motorway!

Railways

As roads, particularly motorways, have improved, Britain's railway system has declined. Reasons for railway decline are:
  (i) cost of track has to be included in the fare, therefore they are expensive
  (ii) expensive over distances of less than c.350 km
  (iii) goods need to be delivered from and to the stations
  (iv) competition from private cars, coaches, buses.

Air

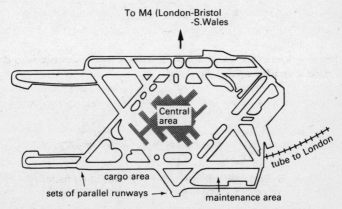

Fig. 100 *Heathrow Airport, built on firm gravel 20 km from central London*

Locational requirements of international airports:
  (i) flat land
  (ii) large area – several long (3 km) runways are needed, to cope with changes in wind direction, and to handle several planes at once
  (iii) well-drained land – runways need firm foundations
  (iv) good access to population centres.

Comparison of transport systems

	Advantages	*Disadvantages*
Road	door-to door service cheap for distances under 300 km	poor bulk carrier pollutes air motorways use up land
Rail	fast safe good bulk carrier low pollution	expensive over short distance needs gentle gradients
Air	very fast can reach remote areas	expensive bulk carrier vulnerable to bad weather noisy

Q. 14 The graph [fig. 101] illustrates a typical relationship between operational costs and distance for certain types of road and rail transport in Britain. Which of the following statements is (are) supported by the graph?
  (1) Liner trains are always the cheapest form of transport.
  (2) For distances less than 300 km, transport by 16-ton lorry is cheaper than by 10-ton rail wagon.
  (3) For distances more than 400 km, rail transport is cheaper than road transport.   [LON B]

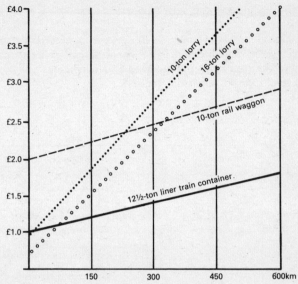

Fig. 101

British Isles: Agriculture

Multiple-choice questions concentrate on testing your understanding of the vocabulary used in, and the factors influencing, agriculture:

British Agriculture: Vocabulary

Arable: field crops

Cereals: grain, e.g. oats, wheat, barley

Crofting: See p. 81

Cultivation: crop production

Extensive: farming with low yields from a large area

Horticulture: market gardening

Intensive: high yield, high **inputs** of labour, fertilizer

Ley grass: grass planted as rotation crop

Loam: fertile soil of mixed sand, clay and humus

Market gardening: vegetables and salads

Mixed: pastoral and arable farming practised together

Orchard: fruit trees

Pastoral: animal rearing

Root crops: turnips, parsnips, swedes, beets

Rotation: alternating crop type to maintain soil fertility

Stock: animals

Q. 15 'There has been a great increase during the past decade in Great Britain in the acreage devoted to X crops, especially to barley which is now the most important.' The word which most accurately replaces X is
A. arable
B. subsistence
C. cereal
D. root
E. rotation

Farming Type	Locational Factors	Sample Areas
Beef cattle	more isolated areas	N.E. Scotland central Ireland
Dairy cattle	*physical* mild winters – over 4°C; cool summers to maintain grass; rainfall over 750 mm, evenly distributed; clay soils	Cornwall N.E. Scotland central Ireland
	human proximity to conurbations	North Cheshire Gower peninsula and Gwent lowlands near London, *although not physically ideal*

Farming Type	Locational Factors	Sample Areas
Cereal production	*physical* warm, sunny summers – over 15°C; rainfall in growing season – c.600 mm; flat land; clay soils	East Anglia Vale of York
Root crops	*agricultural* sugar beet, suitable rotation crops with cereals, potatoes *physical* cooler summers – below 15°C to avoid diseases	Fens Scottish Lowlands
Hops		Kent, Sussex, Hereford and Worcester
Orchard crops	*physical* mild winters rainfall over 750 mm	Cornwall, Devon, Somerset, Hereford and Worcester
Market gardening	*physical* mild winters, early springs *human* demand from cities: London, Midlands, S. Yorkshire, S. Lancs.	Cornwall and Devon Home Counties Vale of Evesham Vale of York Cheshire S. Lancs.

With the aid of your atlas, identify the areas of Grade I farmland shown in fig. 102. Use the geographical direction and the county name: e.g. eastern Norfolk.

Q. 16 (a) Select two of these areas, other than the one marked A. Describe the natural advantages each possesses for farming

(b) What advantages for farming has region A that are not related to soil and climate? [OXF, part]

Essay questions ask about types of farming or about regions. You can cover both types of question by studying three regions which farm in the three most frequently examined types. ('O.D.' means 'above sea level'.)

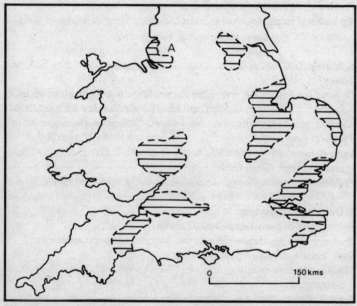

Fig. 102 *Areas of Grade I farmland in England and Wales*

1. Arable farming in East Anglia (Cambridgeshire, S. Lincs., Norfolk, Suffolk)

Relief. Below 200 m except E. Anglian 'heights'; below sea level in Fens.

Drainage. Pumped drainage needed in Fens and S. Lincs.

Soils. Glacial clays and sands. Soil type strongly influences crops grown (below).

Climate. Winters under 4°C, summers over 15°C, rainfall around 600 mm.

Crops
   (i) barley – esp. eastern Norfolk – rich loams
   (ii) wheat – esp. Fenlands
   (iii) sugar beet esp. on Breckland (sandy area) in rotation and with
   fertilizers
   (iv) rotation crops – turnips, potatoes, clover
   (v) peas, beans – this is 'Bird's Eye Country'.
Stock. Cattle – near King's Lynn.
Typical farm. 360 acres; rectangular fields; wheat, sugar beet, vegetables,
   peas, potatoes; drainage ditches; below 100 m O.D.
Problem. Soil erosion in Fens due to removal of hedges.
Agricultural industries. Norwich and Ipswich are major centres for processing the grain and sugar beet of East Anglia.

2. Market gardening in the lower Thames basin (S. Herts, S. Essex, Surrey)

Demand for food is the overriding factor. Should soils be poor, fertilizer
is applied; if climate is too wet and cloudy, glasshouses are used; if rain
is inadequate spray irrigation can be used; fodder crops can be bought
if grass is inadequate.
Typical farm. 25 acres; small fields; vegetables, potatoes; glasshouses
   used. Around 100 m O.D.
Problem. Competing forms of land-use (housing, factories, roads, leisure).

3. Dairying in Somerset

Relief. Plain of Somerset below 100 m O.D.
Drainage. Ditches dug and rivers straightened to improve drainage.
Soils. Fertile alluvium, peat in places.
Climate. Winters over 4°C; summers over 15°C; rainfall 750 mm, evenly
   distributed.
Typical Farm. 200 acres; permanent pasture; ley grass, kale, apple
   orchard; several streams; 200 Friesian cattle.
You may be given a farm plan and asked:
   (i) to describe it
   (ii) to compare it with another.
If you are asked to compare two plans, use a logical order – relief,
drainage, total size, field size, crops, livestock. In comparison questions,
two separate descriptions WILL GET NO MARKS.
   You may also be asked to give a sample location.

Problems in British agriculture

1. Soil erosion (see p. 137).
2. Lack of water. Irrigation is sometimes necessary in areas of less than 600 mm rainfall, especially for market gardening. Sprays are usually used – IRRIGATION CANALS ARE NOT FOUND IN THE U.K.
3. Flooding. In major river basins, e.g. Thames and Trent, in winter. Occurs in flat areas after heavy rain in upper reaches of river. Avoided by embanking rivers, using drainage ditches and **field drains** (pipes laid in fields).
4. Upland farming – difficulties of:
   (i) access
   (ii) gradient – too steep for machinery
   (iii) climate – altitude lowers temperatures, increases rainfall; snow kills thousands of livestock
   (iv) distance from markets
   (v) poor roads.
   Typical upland farm – 650 acres; a few very small fields; most of the acreage is rough pasture; permanent pasture, oats, 450 sheep, 7 dairy cows, steeply sloping land above 350 m O.D.
5. Competition for land (see p. 129).

Recent developments

1. The effects of the **E.E.C. (European Economic Community**, the **Common Market)**. The **Common Agricultural Policy (C.A.P.)** has
   (i) stimulated grain production because of subsidies
   (ii) threatened dairying
                  poultry
                  apple growing
   because of foreign competition. The U.K. orchard owners have banded together to fight back.
2. Changing tastes. We now eat more protein – eggs and meat – and less starch. Consequently livestock production is up; potato consumption slightly down.
3. Irrigation is increasing in S.E. England, as demand for vegetables continues.
4. The agricultural acreage is declining because of new building, etc.; the labour force is declining as mechanization increases.

Q. 17 Select an area on the eastern side of Great Britain where arable farming or market gardening is a leading occupation.

(*a*) *Briefly describe the main features of agriculture in the area chosen.*

[5]

(*b*) *Show how the farming you have described in that area has been influenced by:*

(*1*) *climate* [5]

(*2*) *soil* [2]

(*3*) *communications* [3]

(*4*) *available markets.* [3]

[OXF]

Multiple-choice questions sometimes come in the form of two statements. In both London and the Associated Examining Board examinations, the code letters for a false statement are:

	First statement	Second statement
C	True	False
D	False	True
E	False	False

If both statements are true, answer

    A if the second statement is *a correct explanation* of the first statement

    B if the second statement is *NOT a correct explanation* of the first statement.

The following is an example of this type of question:

Q. 18 A larger percentage of the working population was employed in agriculture in 1958 than in 1978.   *Mechanization on the farms of the U.K. increased between 1958 and 1978.*

[AEB]

NOTE. There is more about Britain's agricultural problems in the section of the next chapter on 'Problems of Developed Countries: British Examples'.

5. World Problems

These are the major topics
1. Problems of population
   (i) population growth
   (ii) urbanization – in the developing world
                     – in the developed world (causes; traffic; urban
                        renewal)
2. Environmental problems
   (i) mining
   (ii) environmental pollution
   (iii) national parks
   (iv) soil erosion
3. Water supply
   (i) drainage
   (ii) drought
4. Problems of tropical agriculture.

The developed world

This is covered in this book by Chapter 4 (British Isles: Human
Geography) and by the sections on
   'Problems of Developed Countries: British Examples' (this chapter)
   'Problems of Developed Countries: Further Examples' (this chapter)
   'Western Europe' (Chapter 8)
   'North America' (Chapter 8).
Questions under this heading tend to be very *vague*: usually no country
is named. It is up to you to be *specific*:
   name places, products;
   give figures for relief and climate;
   draw sketch maps to locate your examples.

The developing world

Some boards ask the generalized type of questions discussed above. Others add some specific questions on named areas. You would be well advised to know the main themes as they apply to one continent, but be aware of similar developments elsewhere. Some problems of the developing world are described in Chapter 6.

Problems of developed countries: British examples

In Britain, like the rest of the **developed world** – Australia, New Zealand, U.S.A., Canada, Japan, Western and Eastern Europe, and Russia – the problems are not so much due to the physical environment as to population pressures and changes.

PROBLEMS OF POPULATION PRESSURE
'Disappearing' countryside
Pollution
Traffic
Water supply

National Parks and open spaces

Like much of Western Europe, Britain is overcrowded. Land is needed for industry, transport systems, houses, schools, hospitals, offices, farming and recreation. Leisure pursuits have become much more important in the last twenty years because of
- population increase
- higher incomes
- shorter working week
- longer holidays
- improved transport (especially increased car ownership).

Britain has planned for leisure since 1945 and the first major development set up nine National Parks (fig. 103).

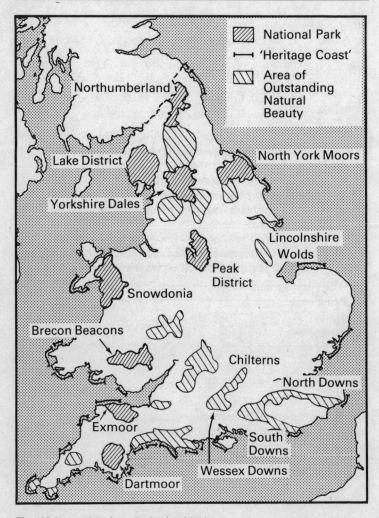

Fig. 103 *National Parks in England and Wales*

National Parks

(1) Aims   To conserve unspoilt country areas and improve public access.

(2) Problems   (i) population density is low and the economy sluggish; farming improvements and industrialization, which would provide jobs, would also disfigure the countryside.

(ii) tourist facilities – camp and caravan sites, car parks, lavatories, gift shops – spoil the beauty tourists want to see.

(3) Some solutions (i) hiding new roads in cuttings
(ii) shielding camp and caravan sites with belts of trees
(iii) buildings constructed out of local stone
(iv) visitors' facilities are concentrated into selected spots so that most of the area is left unspoiled

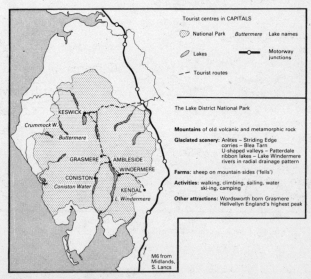

Tourist centres in CAPITALS

National Park   *Buttermere*   Lake names

Lakes   Motorway junctions

Tourist routes

The Lake District National Park

Mountains of old volcanic and metamorphic rock

Glaciated scenery: Arêtes – Striding Edge
corries – Blea Tarn
U-shaped valleys – Patterdale
ribbon lakes – Lake Windermere
rivers in radial drainage pattern

Farms: sheep on mountain sides ('fells')

Activities: walking, climbing, sailing, water ski-ing, camping

Other attractions: Wordsworth born Grasmere
Hellvellyn England's highest peak

KESWICK
Crummock W.
Buttermere
GRASMERE   AMBLESIDE
WINDERMERE
CONISTON
Coniston Water   KENDAL
L. Windermere

M6 from Midlands, S. Lancs

Fig. 104 *The Lake District National Park*

Other 'protected' areas are Country Parks, Areas of Outstanding Natural Beauty, Nature Reserves and National Trust properties – including much Heritage Coast.

NOTE. Several examining boards combine topics as in the following:

Q. 1 Study fig. 103. (Place names not given by examiners.)

(*a*) *Name an Area of Outstanding Natural Beauty and describe its characteristics.*

(*b*) *Name one National Park and*
  (*i*) *describe its attractions*
  (*ii*) *discuss its accessibility from the main conurbations.*

(*c*) *Choose two stretches of Heritage Coast and, by means of sketches and diagrams, explain for each the nature of its interest to the public.*

[OXF, adapted]

You might choose one ria and one stretch of cliffed coastline (Chapter 3).

Pollution

Air is polluted by
  – smoke from coal fires
  – factory fumes, especially sulphur dioxide
  – vehicle exhausts (carbon monoxide; lead).
Water is polluted by
  – industrial waste
  – agricultural effluent and excess fertilizers
  – domestic sewage.
Land is polluted by
  – industrial waste tips
  – coal tips
  – scrap yards
  – domestic rubbish – especially PVC – and litter.
The sea is polluted by
  – oil tanker leaks
  – domestic sewage
  – litter.

Results
  (i) health hazard for humans, e.g. bronchitis. Aberfan spoil tip swamped and killed 116 children and twenty-eight adults
  (ii) death for water species, sea birds, small mammals
  (iii) defacing of buildings by soot, etc.

The solution
To make pollution illegal: but this is not easily done. Conservation is very expensive.

Traffic in towns

Through traffic may be diverted by **by-pass roads** but is not the major problem in large towns where

(i) workers move in and out of the C.B.D. (see p. 85) during rush hours

(ii) lorries and vans deliver goods

(iii) people use the services of the C.B.D.

Solutions	Snags
Pedestrian precincts Multi-storey and underground car parks	construction costs
Widening roads; ring roads	cost of land purchase demolition re-building

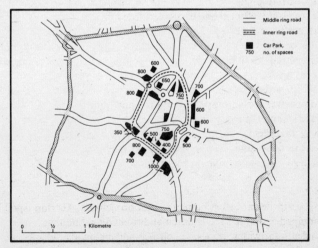

Fig. 105 *Traffic in central Birmingham*

Problems of population change

1. **Rural depopulation**, leaving the countryside, results from
   (i) increased mechanization – fewer jobs
   (ii) decline of 'country' jobs – blacksmiths, thatchers, etc.
   (iii) poor amenities – inadequate public transport, services
2. **Inner city decline**: richer people move to suburbs leaving the poor in decaying housing.
3. **Growth of suburbs** – uses up open space.
4. **Village urbanization.** Villages become dormitory towns. House prices rise to levels local people cannot afford.

Urban renewal

This is one answer to the problems of the conurbations (p. 88). Birmingham plans to develop derelict areas and land cleared of slums to:
   – build better housing, linked with new schools
   – increase available open space
   – develop separate light industrial estates
   – improve the roads.

In addition, new estates have been built on the city outskirts.

Q. 2 (a) Select one major conurbation and describe the factors which have favoured its growth.

The West Midlands (pp. 87–8) would be a good choice.

Q. 2 (b) Explain how problems of
   (i) traffic congestion; (ii) slum clearance and redevelopment; and (iii) declining industries are being solved. [JMB, adapted]

For part b(i), use fig. 105 to describe the ring roads (shape, dimensions) and car parks (location, number, total number of places).

Q. 3 With reference to your studies of the developed world, write an essay on one of the following:
   (a) Pollution: its causes and cures
   (b) Pipelines
   (c) The functions and morphology of one named capital city
   (d) Traffic congestion in towns
   (e) Population movements [LON A]

'With reference to your studies' means 'give named examples'.

You will need to supplement the information from this book with examples in order to tackle (a) or (c). It would be difficult to get enough information for topic (c).

Problems of developed countries: further examples

In addition, the developed world has problems with mining, land reclamation and water supply.

Mining problems

Conditions	Example	Difficulties	Solutions
Arctic	oil: Prudhoe Bay, Alaska	(i) soil surface waterlogged in summer, iron-hard in winter	Specially designed pipeline to Valdez (900 km long)
		(ii) transporting oil	
		(iii) earthquakes	
		(iv) supplying oil crews	Airport construction
Arctic	iron ore: Kiruna, Sweden	Swedish port – Lulea – ice-bound in winter	rail link with Narvik – Norwegian ice-free port
		continuous darkness for five weeks in midwinter	arc lights powered by H.E.P. from Porjus
Mountain	copper: Bingham, Utah	relief – over 2,500 m O.D.	railway links
		distance from markets: West Coast 1,000 km; East Coast 5,000 km	concentration of ore at mine reduces transport costs
			open-cast mining reduces operating costs
Desert	copper: Morenci, Arizona	lack of water: rainfall less than 250 mm	costs of piped water offset by high value of copper

These problems are overcome because world prices justify the expensive solutions.

Land reclamation

Population pressure is so great that land must often be drained, e.g. the English Fenland, the lower Po flood plain of Italy, the marshland (Watten) of the north German coast (F.D.R.). The most spectacular reclamation schemes have taken place in the Netherlands, e.g. Flevoland.

Polder reclamation
(**Polder** is the Dutch word for reclaimed land.)
1. Seal off the area with a **ring-dyke** (embankment).
2. Protect dyke walls from erosion.
3. Face dyke with concrete.
4. Plant dyke with vegetation to prevent wind erosion.
5. Pump polder dry.
6. Leave for four years for rainwater to wash away salt.
7. Deep-plough.

Water supply

The shortage of water for agriculture has been discussed on p. 126. Water is also needed for industrial purposes and demand has risen as industrial development has increased. Pipelines are constructed from the wet upland areas to the areas of demand and water is recycled.

Problems in the developed world and the developing world

Soil erosion

1. Causes of soil erosion

Causes	Examples in the Developed World	Examples in the Developing World
Vegetation clearance removes roots which bind the soil together	forest clearance: American mid-west	burning of savanna woodland when rain forest is cleared, sun scorches soil (northern Ghana)
Downslope ploughing speeds up the flow of rainwater: increased erosion	Appalachian Mts., U.S.A.	Ethiopian Highlands
Monoculture – growing one crop only – exhausts soils	cotton, southern U.S.A.	
Overcropping	Oklahoma: grain	insufficient **fallow** period due to increased food demand (Sierra Leone)
Planting crops in rows and weeding leaves bare strips of soil	corn (maize) in rows resulted in loss of up to three-quarters of the topsoil (Ohio)	coffee in Uganda
Overgrazing – too many animals on the land		common in Africa where cattle are a measure of wealth; aggravated by schemes to bore water-holes and vaccinate cattle, e.g. in Somalia

As well as the problems listed above, the tropical world also has the difficulties of torrential rain after months of drought.

2. Results
1. **Sheet erosion**, where whole areas of topsoil are removed.
2. **Gully erosion** – deep gorges cut in the hillsides.

3. Solutions
(i) To stop water flowing directly downslope by:
    contour ploughing across the slope
    strip cropping of varied crops
    using **cover crops** between row crops (**inter-cropping**)
    planting grass strips
    cutting terraces
    replanting trees.
(ii) To reduce wind erosion by planting wind belts.

Multi-purpose river developments

The first of these developments in the Tennessee Valley, U.S.A., (**T.V.A.**) had many aims, among them the control of soil erosion:

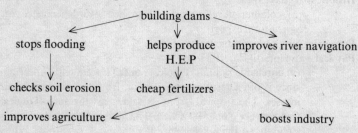

This was backed up by the soil conservation methods outlined above. The T.V.A. scheme was so successful that the idea has been copied in the developing world, e.g. the Volta River Project.

The Volta River Project
In 1962, the construction of the Akosombo dam across the Volta River, in Ghana, began. In 1971, Akosombo produced 99 per cent of Ghana's electricity (2,909 million kWh). In consequence:
    (i) cheap electricity has favoured industrial growth, notably aluminium smelting at Tema
    (ii) Ghana sells electricity to its neighbours
    (iii) Lake Volta is used for transport

(iv) Lake Volta produces a fifth of Ghana's fish

(v) irrigation is a possibility.

But there have been disadvantages. Water-borne diseases (e.g. bilharzia) have increased, and 78,000 people had to be resettled.

Problems of the developing world – Africa, South and Central America, Asia

The overriding problem is to improve living standards (see 'Indicators of Underdevelopment', p. 148).

Causes of poverty

Taking Africa as an example, we may summarize the causes of poverty thus:

Historical (i) slave trade

(ii) colonial development encouraged overdependence on agricultural exports

Economic (iii) unfavourable terms of trade

Physical (iv) infertile laterite soils

(v) unreliable rainfall. **Tropical rainfall** is frequently uncertain since it often results, as in Britain, from the meeting of air masses. The tropical 'front' is known as the **Inter-Tropical Convergence Zone (I.T.C.Z.)**. The converging winds vary in strength, so the convergence may be weak or strong. When it is weak, there is less chance of rain. Furthermore, the I.T.C.Z. moves north and south of the Equator and the extent of its movements varies enormously

(vi) remoteness – distance from sea

– difficult terrain

– unnavigable rivers

– impenetrable jungle

(vii) diseases.

If you are studying the developing world you must learn more about the physical causes of underdevelopment, and about the solutions. You will find Chapter 4 of *Central Africa* by Young and Lowry (Edward Arnold) useful here.

DEVELOPMENT METHODS
1. Improving agriculture (p. 142)
2. Land reform
3. Intermediate technology
4. Reduction of population growth (p. 147)
5. Increasing mineral production
6. Producing cheap power (p. 138)
7. Developing industry – including tourism
8. Improving transport

Q. 4 Fig. 106 shows six physical barriers to development in West Africa.
For each one
    (a) explain what it is
    (b) discuss its causes and effects
    (c) show how this barrier can be overcome.

You should be able to describe and discuss physical barriers to development in this way for your special region within the developing world.

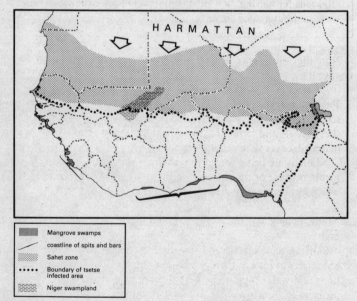

HARMATTAN

	Mangrove swamps
	coastline of spits and bars
	Sahel zone
•••••	Boundary of tsetse infected area
	Niger swampland

Fig. 106

Agriculture in the developing world

First read the short section on 'World Agriculture', p. 190.

Examples of subsistence farming
1. **Shifting cultivation.** This is dying out as population pressure increases. The Zande people of the southern Sudan may be taken as an example:

Arable areas. Fertile valley slopes and forest edge; upland grassland left for hunting

Forest clearings.
　　Use in year 1: millet, groundnuts
　　　　　year 2: maize
　　　　　year 3: cassava
　　　　　year 4: plot abandoned to regain fertility

Fertilizers. Refuse heaps, household waste, ash, added to growing banana, pumpkin and sweet potato crops

Reasons for moving. Soil exhaustion; increasing distance of fields from house; poor harvests

---

Note that subsistence farmers adapt to the environment. They do not change it.

2. **Pastoral nomads** are found in the **Sahel** – the belt of semi-desert and scrub south of the Sahara (fig. 137). The Fulani people raise cattle, sheep, goats and camels for their meat, milk and skins. When pasture in one place is exhausted, they move on.

　　Nomads are being forced to **stabilize** – stay put – by many governments because of the soil erosion they cause by overgrazing (p. 137), which has been aggravated by unreliable rainfall (p. 139).

3. **Multicropping**, e.g. p. 226. You should locate these places on a world map.

　　Shifting cultivators and pastoral nomads had efficient rural economies as long as their population remained low. Modern medications and water supplies have kept more people and stock alive. The problem is, how to use the available resources (**resource management**) without destroying them, i.e. **conservation of resources**.

Some of the answers are found in the soil conservation techniques listed on p. 138. Others lie in improving agricultural output.

Improving agricultural output
1. Using fertilizers and pesticides.
2. Irrigating. *Small schemes* include the use of the Sakia; Shaduf and
   borehole in Egypt.
   Large schemes – probably the most successful in Africa is the Gezira
   scheme in the Sudan (fig. 107).

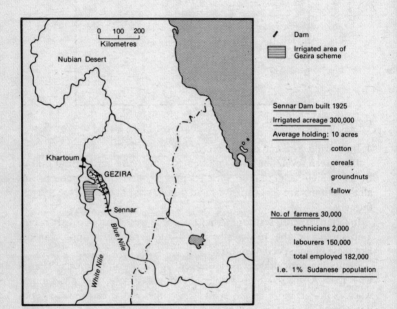

Fig. 107

3. Drainage. There are extensive swamplands in
   the Amazon delta
   south and west Borneo
   eastern Sumatra
   west Africa.

   Drainage schemes are taking place in Sierra Leone (fig. 108). It is
difficult to control the water. Cooperative groups can produce 1,700
lb of *padi* rice per acre – a high yield.

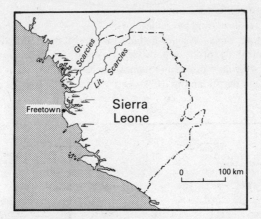

Fig. 108 *The Scarcies estuary, Sierra Leone*

4. Improving crop and animal breeds.

The '**green revolution**' was an attempt to increase agricultural output by using high-yielding crop varieties plus fertilizers and irrigation. But all these things are very expensive and they have not benefited the poorest farmers.

In Ghana disease-resistant 'Amazon' cocoa trees have boosted output. You should be able to describe and locate agricultural improvements like these in one specific country.

Plantation agriculture was developed by European colonists. That is why East Africa, with its cooler, healthier climate, has plantations, whereas hotter, more humid West Africa (the 'white man's grave') does not. Cocoa production in Ghana is unusual in that it is an area of commercial farming introduced and developed by Africans.

Power

Most of the developing world lacks **fossil fuels**: lignite, coal, oil. In Africa, Egypt, Libya, Algeria, Nigeria, Gabon and Angola have OIL; Nigeria and Zimbabwe have COAL.

Power is vital to economic development. Thus, many African countries are anxious to develop H.E.P. You should know the details of one H.E.P. scheme: (i) its importance to the relevant country and (ii) the problems. See Volta River Project, p. 138.

Transport

We have seen how important transport has been to British economic development. Transport is poorly developed in Africa because of:

Physical difficulties – waterfalls on major rivers, e.g. Congo
- rivers drying up, e.g. Niger in Mali
- floods washing away bridges
- dense vegetation
- intense tropical rainfall washing away roads
- sand drifting in deserts

Economic difficulties – cost.

Many African countries have developed air transport to 'leap-frog' the physical barriers. But its cost makes it useless for ordinary people.

Some route centres have developed in West Africa, like Kano, Nigeria, or Kumasi, Ghana. You should be able to draw a sketch map of such a centre – your atlas will help.

Population

Population in the developing world is discussed in Chapter 6.

Industrialization

The colonists tended to develop Third World countries to feed their own industries. Developing countries now have to change from export-dominated economies, where manufactured goods were imported, to manufacturing their own goods.

Industrialization is hampered by:
- lack of power
- inadequate transport
- poor communications
- low educational and technological standards.

Some developing countries like Hong Kong, Taiwan and the Philippines are basing industrialization on cheap labour. Other countries wonder if low wages and long working hours are really what they want.

Individual countries

You should be able to write a **geographical account** (see p. 216) of two countries from either Africa, Latin America or Asia (excluding Japan).

Further Reading
Andrew Reed, *The Developing World*, Bell & Hyman, 1979.

6. Aspects of World Geography

World Population

There are six major topics in this section:
1. Population growth, in the past and at present.
2. The contrasts between the populations of the developed and the developing world. The latter is also known as the less developed world, and is often called the **Third World** by newspapers and television. (For definitions see pp. 129, 139.)
3. The varied age–sex structure of different countries.
4. The variations in population density over the world's land surfaces.
5. The variations in population density within countries.
6. Population movements within countries (**internal migration**).

One examination question might contain aspects of two or three of these themes.

Population growth

Historians have studied the growth in the total population of the world, and find that although population grew very slowly in the past, it began to increase very rapidly by about 1800 and now grows more quickly each year.

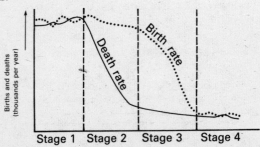

Fig. 109

If we look at fig. 109, which shows Britain's population growth, we can see two things:

(i) population growth depends on the behaviour of **birth rates** – the number of babies born each year per thousand people – and **death rates**, also measured in thousands per year;

(ii) there are several distinct phases, each marked off by the vertical broken lines on the graph. In the first stage, before 1700, birth rates were high, but so were death rates. The total increase was thus small.

In the next stage, agriculture and industry became much more productive, so that there was more wealth, and medicine and hygiene improved, so that the death rate fell. Family sizes remained large, however, so that population growth was rapid.

Later, with increasing wealth and better education, people began to limit their families. The birth rate began to fall since the richer, better-nourished people of the world tend to have fewer babies. (The reasons for this are not fully known, but (i) parents in poorer countries tend to have more children because they need their family to care for them when they are old – there is no old-age pension; (ii) when many children die before their fifth birthday, people have more children in the hope that some will grow up and contribute to the family's livelihood.)

Finally the birth rate is almost as low as the death rate, and population is more or less **stable**.

Britain, like the other developed countries, is in the fourth stage of the graph. Some geographers use this graph as a pattern or '**model**'. They think that regions which now have high birth and death rates – like the Upper Amazon Basin – are in the first stage and must go through all the other stages before the population stops growing.

Question 1 aims to test your understanding of this model: remember that death rates always fall before birth rates.

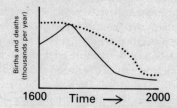

Fig. 110

Q. 1 The graph [fig. 110] shows population trends in the developed world.
  (a) Using the terms birth rate and death rate, label correctly the two lines on the diagram.
  (b) Underline the phrase which provides the most likely explanation for the trends in the second half of the time span shown:
     increasing urbanization
     desire for growing wealth
     higher levels of health and education
     greater mobility of the people.   [OXF]

Sometimes geographers look at the **fertility** rate – births per thousand females rather than per thousand populations. **Less developed countries** (L.D.C.s) usually have high fertility rates.

Some less developed countries do not worry about their rapidly increasing population; in fact they think of their people as their greatest natural resource. The People's Republic of China has used mass supplies of manual labour as a partial substitute for machinery in an effort to overcome poverty. Cuba has a similar attitude.

However, in many less developed countries population growth has become a serious problem. Each year, more and more food must be provided in an effort to feed everyone, and more money is needed for medical and educational services simply to keep standards at existing levels. Extra jobs must continually be found.

India has made great efforts to reduce the number of births each year. A network of family planning clinics has been set up, and everything provided for birth control is free. This has been heavily publicized in the papers and on television and radio.

It is generally agreed that less developed countries must
     control population growth
     increase agricultural output
     earn more by developing manufacturing industries
if they are to avoid serious problems of poverty.

When a country cannot support its people, it is said to be **overpopulated**. It is low living standards that indicate **overpopulation**; we cannot measure it by looking at the **population density** figures. Overpopulation is not the same as **dense population**, as we shall see later in this chapter.

Other ways in which peoples of developed and less developed countries differ are in
   (i) the number of babies who die before they are one year old (**infant mortality rate**) – high in L.D.C.s

(ii) the percentage of the workforce employed in agriculture (very small in developed countries)

(iii) the amount of nourishment people get each day, measured in calories

(iv) the number of people who can read (**literacy rate**) – few people in L.D.C.s can read.

Another indicator is G.N.P. (Gross National Product) per head. This figure is found by dividing the total value of a nation's output by the number of its people. It is a very rough indicator of wealth. The G.N.P. per head in the developed world is four or five times greater than that of the developing world.

The table below summarizes the major differences between the developed and the developing world, and gives some approximate figures for guidance.

	Developed Countries	*Less Developed Countries*
Birth rate	15/1,000	over 30/1,000
Death rate	around 10/1,000 or less	over 15/1,000
Infant mortality rate	below 100/1,000	above 150/1,000
Percentage of workers in agriculture	below 5%	above 50%
Calorie intake per day	nearly 3,000	2,200 or less
Percentage having had secondary schooling	over 80%	under 20%
Literacy rate	over 80%	under 20%

These statistical measurements are used as 'indicators' of underdevelopment. You should be able to say whether a country is in the developed world or in the developing world simply by being given its 'score' in the categories above, even if the examiners do not give you the name of the country. The actual figures for Chad are given below. You need not remember them all, but it would be useful to know one or two concrete examples.

Birth rate 44/1,000
Death rate 23/1,000
Agricultural workers 82%
GNP per head $600 (1980)
Literacy rate 4%

A further contrast between the developed and the developing world is found when we look at the numbers of people of each sex in the various age groupings. Summary diagrams of a population's age/sex distribution are called **population pyramids** and the two shown below come from an O-level question. In part (a) you should comment on the numbers in each case of old people (over 65 years) and young people (under 15 years).

Read off the numbers from the scale given. (IF YOU ARE GIVEN A SCALE IN AN EXAM QUESTION, YOU SHOULD USE IT.) Add male and female numbers to give totals.

Q. 2[Fig. 111] shows the population structures of two contrasting types of countries.

(a) State the main differences in population structure between the developed country and the developing country.

(b) Suggest reasons for the differences you have noted.

(c) What social and economic problems are likely to arise from each of these differing population structures? [OXF]

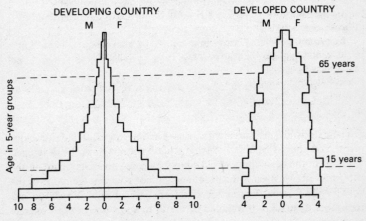

Fig. 111

This section on population growth and the figures on birth rates should tell you why developing countries have so many young people and so few old ones. In part (c) you are asked to note problems arising from *each* of these structures, so indicate clearly which is which. Thus:

'In the developing country nearly 20 per cent of the population is under five and cannot support themselves. It is difficult to feed all these children. In fifteen years these children will be old enough to have children themselves, which will add to the food problem.

'In the developed country there are fewer young people, and these must support the retired people. In the over-65 age groups, there are many more women, often widows, living on their own.'

Questions on population tend to be very straightforward but do depend on your being able to give examples of countries concerned, problems involved and measures taken to solve these.

Population density

The map (fig. 112) shows that some parts of the world's surface are **densely populated**, i.e. there are many people living close together in these areas. Population density is usually given as the number of people living in a square kilometre of territory. Densely populated areas have more than 200 per square km (often written as km^2). Densely populated areas are not always overpopulated. Britain and Sri Lanka have similar, high, population densities – over 210 km^2 – but Sri Lanka's low living standards indicate overpopulation problems.

Moderately populated areas might have around 50 people per km^2. Areas of **low population density** – less than 5 per km^2 – are often called **sparsely populated**.

The map (fig. 112) tries to show the broad differences in living standards ('rich' and 'poor') as well as the distribution of population. (This map shows population distribution by using density shading. Other maps use dots and/or proportional circles.)

You will need to be able to identify some of the countries on this map.

Revision Exercise 1

Make a table like the one shown on p. 152. (You will need the two blank columns for a later exercise.) Using your atlas, name at least *two* countries in each category. You should choose some from the special region you have studied, e.g. Western Europe, North America, the developing world, and some from outside your special region, to improve your general knowledge.

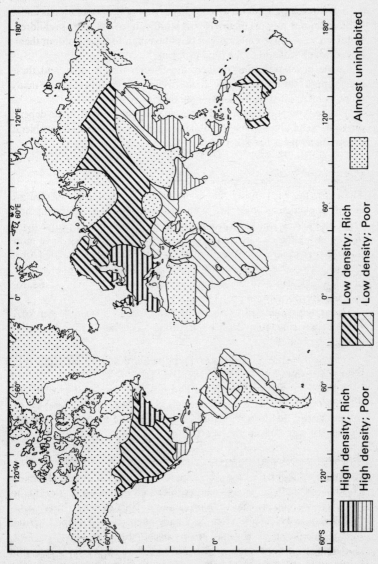

Fig. 112 *World population*

		Country		
Dense population	Rich			
	Poor			
Low population	Rich			
	Poor			
Almost uninhabited				

Revision Exercise 2

Sum up the overall population characteristics of each continent; use the words 'dense' 'moderate' and 'low', thus

Asia – dense

Africa

Australasia

South (Latin) America

North America

Antarctica

Europe

You will realize that the world's population is very far from being evenly distributed.

A key to understanding this is knowing the world-wide variation in farm production. The causes, generally called factors in geography, can be **physical** – to do with natural aspects of the earth – or **human** – concerned with man-made events.

Physical factors in geography

1. **Relief** – the distribution of high and low land. Mountainous land is often difficult to cross (**inaccessible**) and too steeply sloping for ploughing. Uplands have cooler, rainier weather than nearby lowlands. Consequently, people often avoid highland areas, but in some hot, tropical countries, they may prefer the coolness of the uplands.

2. **Drainage.** Rivers may attract farmers because of their irrigation potential or their **alluvial** soils; people in the tropics may avoid running water because it is often a source of disease there.

Try to remember that there are different responses to each factor; don't write as if physical factors completely control people's lives.

3. **Climate.** This is a very important factor in deciding which food to grow. Few people will live where farming is very difficult or impossible: nothing will grow where the temperature is below 6°C; every living thing needs water. This explains the sparse population of the hot and cold deserts, and the ice-caps.

4. **Natural vegetation.** Thick forest can make farming and transport difficult, although most **temperate** forests have been cleared, and the **tropical** forests are being cut down at an alarming rate.

You will see that your knowledge of world climates and vegetation types will be useful in answering questions on population density.

5. **Natural resources.** These include minerals, timber and fish. In the past, minerals like coal used to attract industry and, therefore, people. Oil, on the other hand, is generally moved to where people are already (see p. 208).

The first letters of these physical factors occur, in this frequently used order, in the word ReDisCoVeriNg.

Human factors

There will be more to say about human factors later. Briefly, they can be summarized under four headings.

Economic factors are concerned with buying and selling. Anything man does that changes the amount sold of an industrial or agricultural item has an economic effect. The cost of transport is one such factor, so that developments tend to be in areas which are easily reached. You can see from fig. 112 that there is little population in the cut-off, interior regions of Latin America.

Social factors include the customs or religions of people, which often influence family size, and **political** factors refer to government decisions. Events of the past, like the voyage of the Pilgrim Fathers, are **historical** factors.

Revision Exercise 3
Returning to the table in Exercise 1, head the two blank columns 'Physical Factors' and 'Economic Factors'. In these columns, try to find some of the factors which have brought about the living conditions in the countries you have chosen. Be as specific as you can in naming an important relief feature, type of environment or export crop, for example.

You should now be prepared to tackle the following O-level question:

Q. 3 On the world map [not provided here] ... several areas with a high or low population density are marked.

(a) Define the term density of population.

(b) Referring to at least TWO areas marked on the map, explain how certain economic and physical factors may create a low population density.   [OXF]

In your answer, you should refer to the areas by naming them as accurately as you can. Region A includes Namibia, Angola and Zaire, but if you are not sure about the countries' names, give the part of the continent concerned, i.e. south-central Africa. It is always better to give a more generalized, but correct, answer than a precise, but wrong, answer. The latter will gain no marks at all. Choose the areas you refer to from varied environments; B and C – the Sahara and central Australia – are both deserts, so don't write about both. If your first example was B, go for a contrasting area like D, the southern Andes. Follow the examiner's instructions by giving examples of both physical and economic factors. Then be as specific and concise as you can. In other words, give named examples, and don't be long-winded. Part of your answer might go like this:

'Region D includes parts of Chile and Argentina. One reason for the low population density is the presence of the southern Andes, which rise to over 12,000 feet in places. This makes them too cold for food production, and difficult to travel through.'

This NAMES a location, DESCRIBES the environment with some FIGURES, and EXPLAINS THE EFFECT of this environment. We will see that this pattern is often useful in O-level answers.

Population distribution in a specific country

Population density varies in every country, and this is the basis of the next questions. The regional papers often ask about sparsely and densely populated areas by name (sometimes going across national boundaries to do so):

Q. 4 Account for the very low density of population in Norway and Sweden north of a line joining Bergen, Oslo and Stockholm.   [LON B]

If you go through the physical factors first, in the order found in ReDisCoVeriNg, and then the human – economic, social, political – you are less likely to miss something out. There will not always be something to say for each factor, so you will need to make a plan of which categories to select.

The question is dealt with more fully in Chapter 8.

Q. 5. For one country or region in the less developed world, explain with the aid of a well-labelled sketch map the uneven distribution of population.

[LON A]

Choose a country which is easy to draw and has a simple, clear-cut population pattern. (If you have been supplied with a world map, you may be able to use it to help with the coastline.) Egypt is an excellent choice in this respect (fig. 113). Four categories of population density will be enough. Usually, the denser the population, the thicker and darker the shading. If you follow this simple system the map will look better. You should add the main towns, and insert features you mention in your written answer.

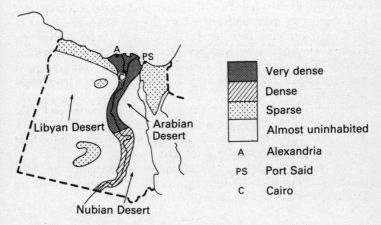

Fig. 113 *Population distribution in Egypt*

A sketch map should ALWAYS be 'well labelled' as this question says, even though these words do not always appear on the paper. If you feel that a sketch map would make your answer clearer, you can draw one even if it is not asked for. Some exam boards always include a blank outline map of each special region. Do use it.

Having identified the different population areas on your map, refer to each type of area in a logical order, i.e. starting with very dense and working down, or vice versa. In each category, NAME the places involved, DESCRIBE the environment with FIGURES if you can and EXPLAIN the effect of this environment.

> **Name – Describe – Explain**
> is a frequent pattern in every area of geography.

One section of the answer might then go like this:

'Except for a coastal strip and two small areas in the west, the almost uninhabited section of Egypt covers all the territory away from the Nile. These areas are all deserts – parts of the Libyan, Arabian and Nubian deserts. They receive less than 250 mm of rain a year, and temperatures are over 25°C for much of the year. Water for farming is only sufficient in the oases. Only a few nomads live in the deserts.'

Urbanization

High population density figures might suggest to you that the people involved are living in towns and cities. In the developed world – Australia and New Zealand, U.S.A., Canada, Japan, Western and Eastern (Communist) Europe and Russia – this is generally so. Population statistics are often used by examiners to bring out this point.

Q. 6 The figures below refer to the population geography of four countries A, B, C and D, in the developed world.

Country	Area in thousands of km²	Population in thousands	Population living in towns of over 100,000 in thousands
A	7647	13013	8209
B	547	52407	21077
C	141	11876	6838
D	9221	22448	12691

(a) Which of the countries appears to be the most highly urbanized?

[LON A]

The examiners are looking for an ability to draw conclusions from groups of figures that they give you, and this ability is being tested more and more these days. To judge **urbanization** – the proportion of people living in towns – you do the following sum:

$$\frac{\text{population in towns larger than 100,000}}{\text{total population}}$$

If you don't have a calculator, round the figures up or down to the nearest thousand, thus:

$$A = \frac{8}{13}; \; B = \frac{21}{52}; \; C = \frac{7}{12}; \; D = \frac{12}{22}$$

You can see that the proportion of urban population in B is clearly less than half, in C and D is just over half, and in A is nearly two thirds. So A is the most highly urbanized.

The question continues . . .

Q. 6 (b) Which of the countries has the lowest mean density of population? This is found by dividing population by area. Again, round the figures up or down, and cancel where helpful:

$$A = \frac{13000}{7000} = \text{nearly 2}; \; B = \frac{52000}{500} = \text{over 100}; \; C = \frac{12000}{100} = 120$$

$$D = \frac{22000}{9000} = \text{between 2 and 3}.$$

The country with the lowest population density should now be clear to you.

In the developing world – most of Africa, Latin and Central America, and Asia – there are highly populated areas which are not urbanized:

Q. 7 Pakistan's population density is more than 100/km², but 23 per cent live in urban centres; Bangladesh's population density is nearly 600/km², but only 6 per cent live in urban centres. How do you explain the fact that Pakistan is more urbanized? [OXF]

Factors encouraging urbanization

1. Changes in agriculture. More machinery (mechanization) or more people (population increase) can mean fewer jobs in agriculture. People might then move to the towns.
2. Industrialization. More industrial output requires more labour; industries tend to group together in towns.
3. More jobs in services. Town people need services like shops, restaurants and offices, and people are needed to run these services.
4. Attraction of towns. Towns have better gas, electricity and water supplies, they are closer to hospitals and schools, and they provide more entertainment.

In your answer to Question 7, you should use the list above and apply it to the two countries named, using place-names, facts and figures of actual examples taken from a regional textbook.

Urbanization is still low in the developing world, due to **internal migration**. African societies, for example, were not based on town life. Before the European colonists came, there were few African towns. One exception is in West Africa, particularly in Ghana and Nigeria. If you compare the plan of Ibadan (fig. 114), an African-developed town in Nigeria, with the plan of Dakar (fig. 115), developed by Europeans in Senegambia, you will see that the two types of town look very different.

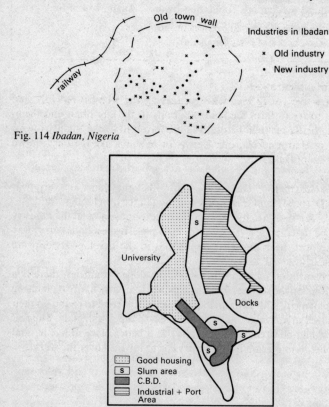

Fig. 114 *Ibadan, Nigeria*

Fig. 115 *Dakar, Senegambia*

The cities of the less developed countries are, however, growing fast as people move away from the countryside.

PUSH AND PULL IN THE DEVELOPING WORLD

Migrants (anyone who moves house) are 'pushed' away from agriculture by
    (i) increasing population pressure leading to lack of land and jobs
    (ii) poor harvests and insufficient food.
'Pulled' to the cities
    (i) better services
    (ii) the hope of a job in services or industries
    (iii) the chance of better housing.

Effects of urbanization

Often it is the young men who go, leaving the women, old folk and children to try to farm the land. Migrants to the city often find themselves no better off than before, and are forced to live in 'shanty towns' like the ones around Nairobi, Kenya, or in slums like the ones shown on the plan of Dakar.

The sketch map of Dakar also shows that the slums are very near the **C.B.D.** (p. 85), and this is very common in the developing world.

Notice also that the city falls into very clear-cut zones; all the industry is grouped together near the port, for example. The best houses are away from the port, near the university. Cities in the developed world are not as clearly 'zoned' as this.

As urbanization increases in the less developed countries, there is a severe strain on the services the cities can provide, particularly on housing and employment. The Tanzanian government has tried to round up their unemployed urban migrants to return them to the countryside. The real answer would be to make the countryside a better place to live in, and this applies to migration from countryside to town all over the world.

City sizes

It is possible to make a 'league table' of cities, by creating size groupings, e.g. over 1,000,000

   750,000–1,000,000
   500,000–750,000
   250,000–500,000
   100,000–250,000

and counting the number of cities of one particular country which fall into each group. This table is called an **urban hierarchy.** (A **hierarchy** is a system of different ranks, one class above another.)

Some geographers believe that developed countries have urban hierarchy patterns different from those of less-developed countries. The developed country is said to have one very large city, two or three in the next group down (rank) and more and more as you go down the ranks. This idea can be shown in the form of a graph, as in Question 8.

Q. 8 The largest city in a country contains one million people. Using the graph [fig. 116] to help you, underline the most probable number of people living in the fourth *city in that country:*

A 750,000; B 600,000; C 500,000; D 350,000; E 250,000.

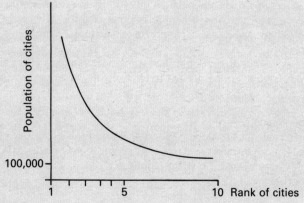

Fig. 116

To answer this question you should construct a vertical line from the coordinate you are given on the *x* axis, i.e. from 4, to the curve on the graph. Now draw a horizontal line from the point you hit the curve back to the *y* axis. This will give you the answer (see fig. 117).

It is always sensible to draw these coordinate lines carefully, at right angles with a ruler, rather than guessing or drawing 'by eye'.

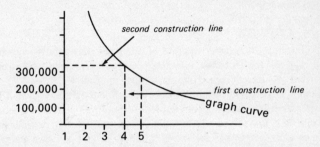

Fig. 117

This idea of urban hierarchies is another example of a **model** – the picture the geographer makes to try to sum up the way things are. The four stages of Britain's population growth, given earlier, are used as a model to predict what will happen in other countries. Remember that models show an idealized picture of a situation. Real life is more complicated.

In less developed countries, things are more extreme. There may be one very large city (the **Primate City**) but very few in the next size groupings.

Further Reading
If you include the Developing World or World Problems as one of your special topics, you might read sections 5 (Population Growth), 6 (Population Movement) and 7 (Urbanization) in *Tropical Lands* by Michael Senior (Longman, 1979).

World Relief

Fig. 118 (over) shows some aspects of the world's **structure**: how it has been made.

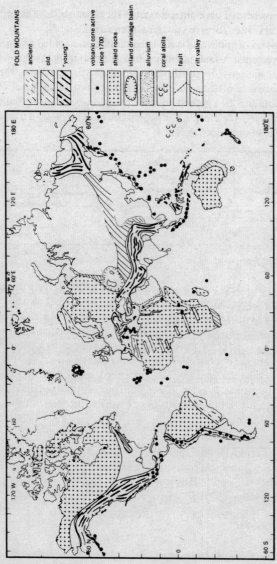

Fig. 118 *World relief*

Shields consist of very ancient, stable rocks which have been heavily eroded. They often form flat uplands: **plateaux**.

Many parts of the world have been faulted and folded as a result of earth movements. A **fault** is a sharp break through rocks where a block on one side has moved – up, down or sideways – in relation to the block on the other side of the fault. Fig. 119 illustrates the different features made by faulting. In a **normal** fault the block slides down the fault slope; in a **reverse** fault the block is pushed up the fault slope; in a **transcurrent** fault the movement is sideways.

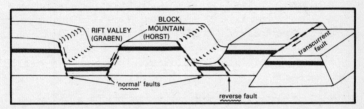

Fig. 119 *Some features of faulting*

Folding has taken place at different stages in the earth's history. We call the most recent fold mountains 'young' even though they are 65 million years old. The older the fold mountains, the lower their height. Fold mountains and volcanoes are thought to result often when sections of the earth's hard crust – **plates** – are pushed about by molten magma rising up from the interior. Fold mountains are the wrinkling of the crust when two plates collide. Upfolds are called **anticlines** and downfolds **synclines**. Immediately after folding, anticlines form uplands and synclines form valleys, but after prolonged erosion the effect of the folding may be lost (fig. 120).

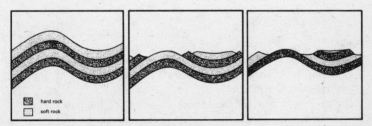

Fig. 120 *Stages in the erosion of folded rock*

Volcanic activity

Notice on fig. 118 that many young fold mountains are associated with **volcanicity** (the formation of volcanoes), as they both form where plates collide. At this point, one plate sinks beneath the other and 'melts', releasing magma. When the magma appears above the surface it is called **lava**, and is a form of **extrusive** igneous activity (p. 51).

Cone volcanoes

A **cone volcano** is a lava-built structure. Lava can vary in thickness (**viscosity**) from solid lumps and ash to liquid. It often contains explosive gas. The shape of the cone depends on how viscous and gassy the lava is. Often lava and ash are interlayered, forming a **composite volcano** (fig.121). Sometimes a volcano blows its top off, leaving an enormous hole called a **caldera**.

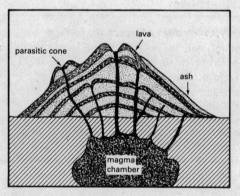

Fig. 121 *Composite volcano*

The effects of volcanoes

Good

    (i) a light fall of volcanic ash can make fertile soil, e.g. Sunset Crater's explosion in Arizona made a moisture-holding soil used by Indians to grow corn for centuries

    (ii) trees grow well on the **extinct** volcanoes of the San Francisco mountains

    (iii) volcanoes are often tourist attractions, e.g. Vesuvius, Fujiyama.

Bad
> (i) clouds of gas and ash can cause appalling destruction; in 1902, one such cloud in Martinique killed 30,000 people
> (ii) Explosions cause (1) earthquakes (**seismic activity**)
>   (2) seismic sea waves
> 75 cubic kilometres of rock were blown off the top of Krakatoa in 1883 and thousands were drowned in the sea wave (**tsunami**) which resulted.

Basalt eruptions

When magma rises beneath the **continental crust**, it can melt the plate, emerge, and flood thousands of square kilometres with basalt lava. These are called **flood basalts**. If magma breaches the **ocean crust** in this way it forms a **shield volcano** which may be built high enough to rise above sea level as a volcanic island. Shield volcanoes are gently sloping, with broad tops, e.g. Hawaiian Islands.

Igneous activity which does not reach the surface

In an **intrusion** the magma cools slowly and large crystals separate out. Later, erosion may reveal these crystalline rocks (fig. 122):

> **Batholith:** large intrusion: eroded to form upland
> **Boss:**   smaller version of batholith
> **Sill:**   flat sheet of igneous rock
> **Dyke:**   vertical sheet of igneous rock.

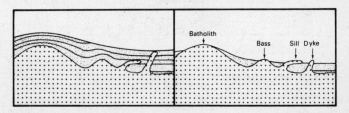

Fig. 122 *Igneous rocks before and after erosion*

Alluvial plains

Alluvial plains are large food-plain areas (p. 63) covered in stream-laid sediments. Alluvium is generally very fertile. **Deltas** form where rivers drop sediment into lakes and seas, and are either:

(i) **arcuate** – having a curved shoreline and many **distributaries** (fig. 123)

(ii) **bird-foot** – with long 'fingers' of silt along the distributaries (fig. 124).

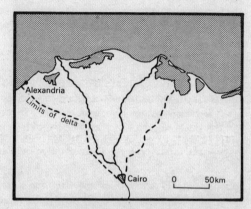

Fig. 123 *Arcuate delta – the River Nile*

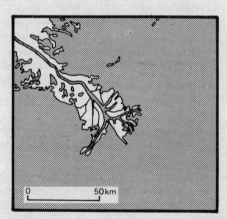

Fig. 124 *Bird's-foot delta – the River Mississippi*

The extensive, flat areas of deltas have supported large agricultural populations, and important cities are often situated on, or near, deltas, e.g. Alexandria, Egypt.

Further deposition may leave these port-cities far from the sea and then it is difficult to keep the river channels open for ships to reach the port. Special jetties can be built to maintain the channels.

Inland drainage basins

Not all rivers reach the sea. Some flow into lakes like the Caspian Sea.

Coral

Coral is made by a tiny marine animal which secretes lime. Atolls are ring-shaped coral reefs, e.g. Bikini Atoll in the Pacific Ocean.

Processes and landforms in hot deserts

1. Weathering

Daytime temperatures in deserts may rise to 40°C, but heat radiates out during the cloudless nights and temperatures may fall to 0°C at night. This large **diurnal** (daily) temperature range causes much expansion and contraction of the rocks, and is thought to crack them up (**onoin weathering**). This mechanical weathering process seems to be effective only when there is some moisture present.

2. Erosion by wind

The wind picks up large quantities of weathered material: sand. Three processes are important:

(i) **deflation**: the removal of material by the wind; this can leave very large **deflation hollows** like the Qattara depression in Egypt

(ii) **abrasion**: sandblasting. This affects the lower rocks, and can
   – cut gorges with ridges (*yardangs*) between
   – cut mushroom rocks (*zeugen*)
   – polish pebbles

(iii) **attrition**: the rounding of the sand grains as they jostle together in the sand storms.

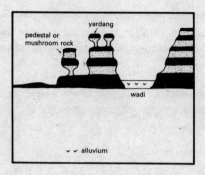

Fig. 125

3. Deposition by wind

The wind drops the sand in the form of long ridges: **seif dunes** and crescents – **barchans**. The barchans are constantly re-formed and creep slowly forward (fig. 126).

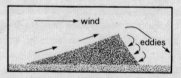

Fig. 126 *Cross-section of a barchan*

4. Water action

The presence of steep, rocky-sided, flat-floored valleys – **wadis** – leads some geographers to suppose that desert landforms may have formed during earlier, wetter times as these valleys are now dry. Others think that although desert rainfall is low – under 250 mm – the rare, heavy rainstorms are responsible for wadis.

The features described in this World Landforms section are tested by the

Oxford board on a world map exercise. Whatever board you take, you will need to know where various landforms occur, together with names.

Revision Exercise 4

Use fig. 118 and your atlas; on a world outline map, locate and name:
  two shield areas
  one rift valley
  one young fold mountain area containing a named active volcano
  one old fold mountain area with no volcanicity
  one alluvial plain
  two deltas
  one inland drainage basin
  one area where coral atolls are situated
  one (desert) area where wind erosion is important.
(Based on four Oxford questions.)

Otherwise you may find 'sectioned' questions:

Q. 9 (a) Recognize features from a photograph.
  (b) Explain how the diurnal temperature range affects desert rocks.
  (c) Draw a fully labelled diagram to explain the features of a barchan.
  (d) State three differences between a yardang *and a* zeugen.

[CAM, part question]

Essay-type questions are less helpful. You must *always* illustrate landform questions, i.e. use fully labelled diagrams, even when this is not specified. Try to use a logical order, like that used in the section on desert landforms on pp. 167–8. Break up the writing into paragraphs: one point per paragraph. Don't worry if this makes very small paragraphs. A typical question is:

Q. 10 Write an essay on O N E of the following:
  volcanic landforms; weathering and erosion in hot deserts; glacial activity; karst landscapes. [LON A]

You can see that you will also need Chapter 3 to cover world landforms fully.

World Climates

The climates of the world are many and complex: fig. 127 gives a simplified picture of the ten most often questioned climatic regions.

A climatic region has distinct features of temperature and rainfall which form a more or less regular pattern throughout the year. We will look at the amount and annual distribution of rainfall, summer temperatures, winter temperatures and the **temperature range**, i.e. the difference between highest average summer and lowest average winter temperatures.

What Climate Figures Mean

Temperatures	extremely cold	below −10°C
	very cold	−10°C
	cold	0–5°C
	cool	5–10°C
	mild	10–15°C
	warm	15–20°C
	very warm	20–25°C
	hot	25–30°C
	very hot	over 30°C

Temperature Range		
small	*moderate*	*extreme*
0–10 deg.	10–20 deg.	over 20 deg.

Rainfall: Totals			*Rainfall: Annual Distribution*		
heavy over 1,500 mm	moderate 500 − 1,500 mm	light under 500 mm	summer maximum	winter maximum	(even) all-year -round

The ten major regions listed in the table on pp. 172–3 all have different combinations of these features. Sample figures are given for each region, but it is more important for you to have an idea of the meaning of the figures (e.g. warm, heavy rainfall) than to learn them parrot-fashion.

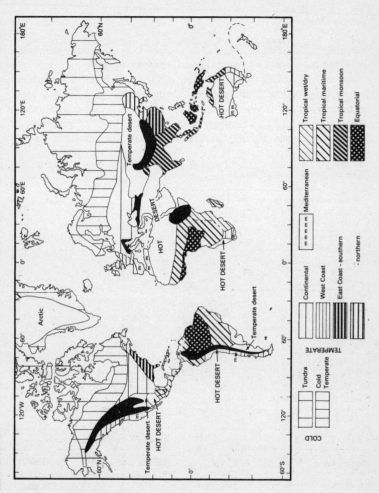

Fig. 127 *World climates*

Climatic Region	Temperature			Rainfall	annual distr
	summer	*winter*	*range*	*total*	
COLD CLIMATES					
I Tundra	mild	extremely cold	extreme	light	ever
II Cold Temperate	warm	very cold	extreme	moderate	sum ma
TEMPERATE CLIMATES					
III Continental	very warm	very cold	extreme	moderate	sum ma
IV Temperate West Coast	warm	cold	moderate	moderate	ever
Va Temperate East Coast	very warm	cold	moderate	moderate	sum ma
Vb Warm Temperate East Coast (as above, but warmer, wetter)	hot	cold	moderate	heavy	sum ma
VI Mediterranean	hot	mild	moderate	moderate	win ma
TROPICAL CLIMATES					
VII Tropical Maritime	hot	warm	small	moderate	even
VIII Tropical Wet-Dry	very hot	very warm	small	moderate	sum rai wi dr
IX Tropical Monsoon	very warm	very warm	small	heavy	mon rai
X EQUATORIAL	hot	hot	small	heavy	even

Main Influences on Climate	Example					Other Locations
	name	summer	winter	range	rainfall total	
itude	Dawson City (Canada)	14°C	−30°C	44 deg.	315 mm	northern U.S.S.R., Greenland Coast
itude	Winnipeg (Canada)	18°C	−15°C	33 deg.	530 mm	north central U.S.S.R.
latitude; ce from sea	Denver (U.S.A.)	24°C	−5°C	29 deg.	510 mm	Russian steppes, S. African Veld
latitude; sea currents	Oxford (U.K.)	18°C	0°C	18 deg.	630 mm	British Columbia, southern Chile, New Zealand
latitude	Washington (U.S.A.)	25°C	1°C	24 deg.	1,064 mm	N.E. U.S.A., N. China
ical latitudes	Hankow (China)	25°C	6°C	19 deg.	1,300 mm	S.E. U.S.A., S. China
ical latitudes, sions in winter	Marseilles (France)	26°C	10°C	16 deg.	580 mm	southern California, central Chile, southern tips Africa and Australia
tudes, ating effect of sea	Rio de Janeiro (Brazil)	28°C	24°C	4 deg.	110 mm	east coast Africa
tudes, ad sun brings rains	Zungern (Nigeria)	34°C	26°C	8 deg.	1,100 mm	central Brazil
tudes, intense heating o low pressure s	Calcutta (India)	25°C	21°C	4 deg.	1,600 mm	N. Indonesia, N. Australia
r − equal lengths of d night, no seasons	Entebbe (Uganda)	24°C	23°C	1 deg.	1,500 mm	Amazon, New Guinea

The easiest questions give you climatic statistics, and ask for a location.

Q. 11 From the world map, choose one area where . . . the following climatic details are true:

January 25°C, July 18°C; annual rainfall 1,600 mm, with 900 mm from November to April.   [OXF]

First think what the temperature figures mean: January (very warm) is warmer than July (warm). If summer is in January, we are in the Southern Hemisphere. Rainfall is heavy with a summer maximum. This is *most like* type IX, in the Southern Hemisphere. Refer to fig. 127 for a suitable location. Do not expect figures in the examination paper to be identical to those in the table.

WARNING. Some boards give rainfall figures in centimetres (cm). Look for the units: add 0 to change from cm to mm, if necessary.

Other world climate questions demand knowledge found in Chapter 2.

Q. 12 A map shows areas of precipitation (i) under 250 mm, (ii) over 1,500 mm.

(a) Briefly explain how mean annual rainfall figures are gathered.

You might be asked the same question, but about average monthly temperatures. In each case

(i) name the appropriate instrument (p. 35)

(ii) explain that recordings are made every day and that mean = average. Thus

$$\frac{\text{Total for all daily recordings}}{\text{Number of daily recordings}} = \text{AVERAGE}$$

The figures can be taken over a month (mean monthly) or over a year (mean annual).

Q. 12 (b) Briefly describe three other aspects of precipitation information that the mean annual precipitation figures do not show.   [LON]

These are:

distribution throughout the year

reliability of the precipitation

type of precipitation (rain, snow, etc.).

Instead of figures you might be given several graphs and asked for likely locations. Again, translate the figures into adjectives for

highest summer temperature

lowest winter temperature (note whether this is in Nov., Dec., Jan., Feb. = NORTHERN HEMISPHERE; May, June, July, August = SOUTHERN HEMISPHERE

range (subtract summer high figure from winter low)

total rainfall

rainfall distribution.

Use the tables on pp. 172–3 to help you with one such question:

Q. 13 (a) The three graphs [fig. 128] represent climate stations in [different] regions ... Link ... a region with the appropriate graphs A, B or C.

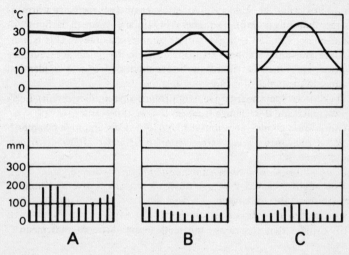

Fig. 128

You should do the description in rough first. Then you will be well placed to answer the next part:

Q. 13 (b) For each region, describe

    (i) the features of the temperature and rainfall distribution which make it different from the others

    (ii) the time of year when man is faced with problems caused by the climate. [SCOT]

You should be able to explain the chief features of a climatic region: that is, outline the main influences on it.

The main factors influencing climate (fig. 131)

1. Latitude

    – The nearer a location is to the equator (low latitudes), the more intense the heating effect of the sun.

    – The equator has overhead sun most frequently and therefore has very

little seasonal temperature variation. Other areas within the tropics have overhead sun during one season.
– Tropical rains are associated with the overhead sun.
– Temperate latitudes (45°–55°) do not get overhead sun, and have clearly marked seasons.
– In high latitudes (over 55°) even the long summer days don't get very warm because the angle of the sun's rays is so low.

2. The sea

(i) Coastal, 'maritime' areas are more moderate in temperature and generally wetter than inland, 'continental' areas. Contrast Continental III and Temperate West Coast IV and Tropical Maritime VII with Tropical Wet-Dry VIII.

(ii) Ocean currents warm up some areas in winter. Contrast Temperate West Coast with Temperate East Coast. Cold currents often help deserts to form (fig. 129).

3. Altitude

Temperatures fall with height, so the main mountain ranges are not included in maps of climatic regions. At the equator, you might pass through a number of climatic zones as you climb a mountain (fig. 130).

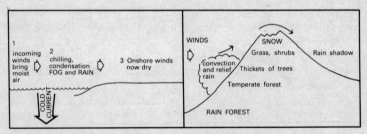

Fig. 129 *How a cold current can help to form a desert*

Fig. 130 *The climatic zones of a mountain at the Equator*

Climatic hazards

Hurricanes start when one section of air becomes very hot. This could

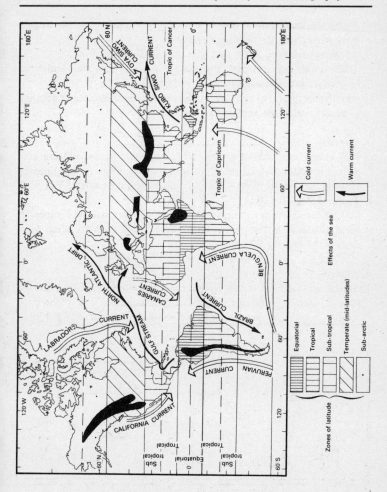

Fig. 131 *Factors affecting climate*

be an island, heating more rapidly than the surrounding sea. The intense heating causes very low pressure. Since:

   (i) 'Winds blow

      From high to low', and

   (ii) winds blow almost parallel to the isobars,

a very strong system of revolving winds with heavy rain results, causing destruction on land and danger at sea.

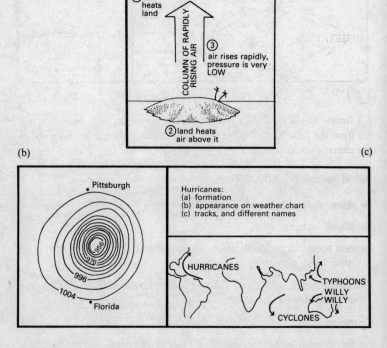

Fig. 132 *Hurricanes: (a) Formation; (b) Appearance on a weather chart; (c) Tracks, and different names*

Drought is a particular problem in regions VIII, Tropical Wet-Dry, and IX, Tropical Monsoon, where rains often fail. Mali, W. Africa, had seven years of drought. (See p. 139.) Death of animals, crop failure and starvation result. Storage of water in reservoirs seems to be the only answer.

World Vegetation

Natural vegetation is the plant life which has developed in any area without man's intervention. Vegetation is very much interrelated with other physical factors in geography.

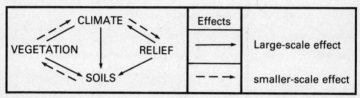

Fig. 133

Climate has a particularly strong effect on vegetation, and you will see from the table on pp. 182-3 that most vegetation types are adapted to the climatic conditions in which they live.

World vegetation: Vocabulary

air plant: see 'epiphyte' (also called **bloom mat**)

broadleaf: wide, flat leaf

buttress roots: large roots, many joining trunks 3 m above ground to support very high trees

Chaparral: scrub and dwarf forest in 'Mediterranean' zone of central and S. California

chernozem: dark, fertile soil, rich in grass roots and humus

conifer: Cone-bearing tree, usually evergreen and needle-leaved

drip tip: exaggerated point to waxy leaf; channels rain water off quickly

dwarf tree: less than 3 m high (also known as **stunted**)

epiphyte: plants living out of contact with soil; usually growing on limbs of trees

evergreen: tree or shrub that holds most of its green leaves throughout the year

liana: woody vine supported by trunk or branches of tree

Maquis: as 'Chaparral', but in southern France

needle leaf: very narrow leaf, emitting little moisture

Pampas: grasslands of Argentina and Uruguay

podsol: infertile soil with hard iron layer

Prairies: grasslands of central southern Canada

raingreen: sudden growth of green shoots after rain

redwood: very large evergreen tree found in northern California

sclerophyll: hard-leaved evergreen trees and shrubs capable of enduring long, dry summer

scrub: low, scattered plant covering

shrub: low, woody plant

Taiga: Russian name for cold coniferous forest

tall tree: 15–30 m high (very tall, about 50 m)

tap root: long root for areas where water is at depth

Tundra: arctic area with eight or more months of low temperatures; subsoil permanently frozen

tussocks: isolated clumps of grass

undergrowth: plant growth below tree layer

Veld: grassland of South Africa

xerophytic: adapted to a dry environment

Natural vegetation: syllabus requirements

Most syllabuses include natural vegetation, but the levels of knowledge required vary. We will start with the most basic, level 1.

Level 1. Knowledge of the main vegetation types and their location on a world map

Use an atlas, the glossary above, the table (p. 182) and fig. 134 to answer Question 14.

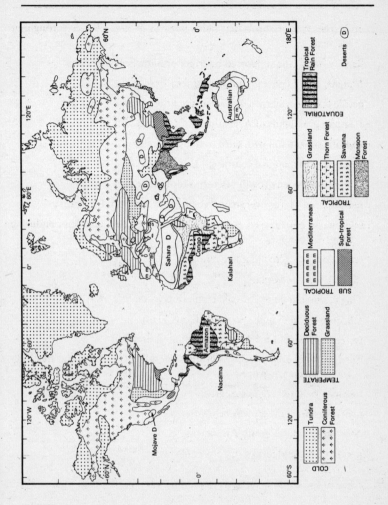

Fig. 134 *Some important vegetation areas*

A. *Vegetation Type*	B. *Other Names*	C. *Characteristics*	D. *Adaptation to Environment*
I TUNDRA		dwarf trees (willow, birch), grasses, mosses, lichens	shallow roots due to permafrost; dwarf trees due to low available moisture
II CONIFEROUS FOREST	Taiga needle-leaf evergreen	straight-trunked conifers, needle-leaved: larch, fir, redwood (softwoods); few species, little undergrowth	leaves adapted to retain moisture
III TEMPERATE FOREST	broadleaved deciduous	low shrub layer grows vigorously in spring	winter leaf-fall conserves moisture
IV TEMPERATE GRASSLAND	Steppe	tall grasses, flowering in summer	grasses die in winter; too little moisture for trees
V 'MEDITER-RANEAN'	Maquis	evergreen trees, broad-leaved: cork oak, Aleppo pine; eucalyptus, acacia in S. hemisphere chaparral (U.S.A.)	trees adapted to long summer drought by ever-green leaves, long tap root down to water table
VI SUB-TROPICAL FOREST	broad-leaved evergreen	trees lower than temperate forest, leathery leaves: evergreen oak, laurels, southern pines, bamboo	leathery leaves retain water
VIIa SAVANNA*	savanna woodland	trees medium height, grassland, park-like appearance; 'raingreening'	xerophytic trees with small leaves, thorns; fire-resistant species only
VIIb THORN FOREST*		elephant grass – up to 3 m high – thorn trees; more sparse than savanna	
VIII MONSOON FOREST		'open' forest, abundant undergrowth	deciduous trees adapted to long dry season
IXa TROPICAL RAIN FOREST	selva	tall deciduous broad-leaved trees in 3 height layers – 50 m, about 15 m, about 10 m – many species	buttress roots support tall trees
	EQUA-TORIAL FOREST (subdivision)	epiphytes, lianas, no leaf-fall season, hardwoods (mahogany, teak, sapele)	drip tips and waxy leaves channel rain

* Sometimes classed together.

E. Climate				F. Soils	G. Location Examples
Summers	*Winters*	*Rainfall totals*	*Rainfall distribution*		
cool	very cold	light	even	Tundra	N. Canada N. U.S.S.R.
mild	cold	moderate	summer max.	Podsol	Canada U.S.S.R
mild	cool–cold	moderate	even	Brown Forest	Britain W. Europe
warm	cold	moderate	summer max.	Chernozem	Pampas Veld
warm–hot	mild	moderate	winter max.	↑	
warm–hot	mild	moderate	even		S.E. U.S.A. S. China Japan
hot	warm	light	summer max.	Laterites common	Sudan
hot	warm	light	very brief summer rians		Tanzania
hot	warm	heavy	summer max; very marked wet season		Sumatra
hot	hot	heavy	brief dry season		Southern Ghana
			even	↓	Congo Amazon

Q. 14 Locate one area of:
  (*a*) *broad-leaved deciduous forest with mulberry and dwarf bamboo*
  (*b*) *evergreen trees with xerophilous bulbous plants, interspersed with maquis scrubland*
  (*c*) *dwarf, shallow-rooted shrubs, e.g. bilberry, carpet mosses and lichens*
  (*d*) *sparse, deep-rooted thorny shrubs, with small, waxy leaves.*

[OXF, from four papers]

Note that the descriptions do not exactly match those in the table (p. 182). They will rarely do so; you must learn to isolate the most important characteristics of each vegetation type.

Level 2. Explanation of vegetation found
This *must* include a knowledge of the climate of each vegetation area and the adaptation of the vegetation to the environment.

Q. 15 The following is a list of some of the characteristics of three natural vegetation types:
  broad-leaved trees; long tap roots; tall tussocky grass; buttress roots; lichens; lianas; stunted trees; scattered trees; bloom mats.
  (*a*) *After careful study of the list, name the three types of vegetation represented and list the correct characteristics of each.*
  (*b*) *Select two of the natural vegetation types you have named in (a) and account for their formation.*   [LONA]

Level 3. Seasonal variations
With the exception of tropical rain forest, most vegetation varies throughout the year. You will need more detail than that given here to answer the last part of

Q. 16 [for savanna, tropical monsoon forest and tundra] explain why all three vegetation types display marked seasonal variations.

[part question]

Level 4. Comparisons
If you decide to tackle 'natural vegetation' you must tackle all the types in the table on p. 182, because examiners often ask for comparisons, usually between types of (i) forest and (ii) grassland.

Revision Exercise 5 (using the table on pp. 182–3)
Copy out and complete the table:

		Similarities	*Differences*	*Reasons*
'DWARF' TREE AREAS	Tundra, I Mediterranean, V Thorn forests, VIIb			
FORESTS	Coniferous forests, II Tropical rain forests, IX			
GRASS-LANDS	Savanna (tropical grass- land, VIIa) Temperate grassland, IV			

This will help you with questions like:

Q. 17 How do the features of the taiga differ from those of the selva?
Describe the soils associated with each type.
Compare the forest products of each type. [LON]

You may be asked to recognize vegetation types from photographs
(Level 4). If this is the case, you should familiarize yourself with the
appearance of the major vegetation types (notably types I, II, IV, V,
VII, VIII and IX) and be able to describe the photographs, using the
table on pp. 182–3 (column C). Thus:

Q. 18 Describe the features of the vegetation in the photograph. Name,
and locate on a sketch map, *where the photograph might have been taken.*
[CAM, part question]

A rarely tested level of knowledge is exemplified in

Q. 19 With the help of specific examples, show how man has adapted
his use of the land to the climate and vegetation of a savanna area.
[AEB, part question]

The Associated Examining Board favours this type of question. If you
are taking this board, choose an area of savanna and consult a regional
textbook to answer this, but remember that only six marks are given
for the answer. AEB candidates might also do the same for vegetation
types IV, V and IX.

PROBLEMS ASSOCIATED WITH NATURAL VEGETATION

There are now very few areas where vegetation is 'natural'. Man has changed plant communities by forest clearance, especially in temperate forests and increasingly in the rain forest; by burning, in African rain forests and savannas; and by deliberate planting of different species, for agriculture and for forestry.

Thus it is that in the African savanna only fireproof species remain. Man's interference here has been dangerous: burning and over-grazing (p. 137) lead to **soil erosion**.

Clearance of the tropical rainforests of the Amazon and Congo may be destroying huge natural reservoirs (for trees hold vast quantities of water). This could lead to a change in world climates. Experts worry increasingly about **conservation**.

World Resources

Minerals

Minerals are often divided into
   ferrous metals – iron
   non-ferrous metals
   non-metallic minerals e.g. phosphates, nitrates
   fuels: coal, oil, natural gas
   water
Details of the methods of mineral extraction and refining are required by few boards, notably the Joint Matriculation Board. For this board you should know
   (i) the world distribution
   (ii) the mining method
   (iii) the refining method
   (iv) the export markets
   (v) uses
   (vi) industrial developments associated with ... aluminium, copper and iron ore.
Other questions relate to the difficulties of mining (p. 135).

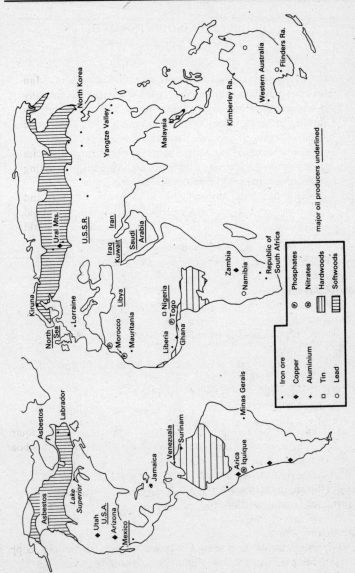

Fig. 135 *World resources*

Minerals are often of great importance in the less developed world, as the following table illustrates.

	Mineral	*Percentage of exports by value*
Libya	oil	99
Nigeria	oil	92
Mauritania	iron ore	52
Liberia	iron ore	77
Zaire	copper	60
Zambia	copper	94
Guinea	bauxite	60

Mineral development has paid for transport developments in Liberia, Nigeria and Zambia; Algeria, Nigeria and Ghana are developing steelworks.

But these are **non-renewable** resources – when the mines are exhausted there will be no more. It is very dangerous to base an economy on such a resource. Algeria – which produces oil – is trying to use the revenue to diversify industry.

Other problems include competition, high transport costs and world price fluctuations (copper was £1,400 per tonne in April 1974; by December the price had fallen to £500 per tonne). Oil prices are constantly in the news because

(i) the developed world is so dependent on it for transport, heating and as a raw material

(ii) the developed world is trying to cut back on oil use

(iii) some **O.P.E.C.** countries (belonging to the Organization of Petroleum Exporting Countries), like Nigeria and Mexico, are heavily dependent on oil revenues.

Timber

Most of the world's timber resources come from the coniferous forests (**soft wood**) and tropical rain forests (**hard wood**).

Coniferous forests
Coniferous forests are easy to exploit because

(i) species have straight trunks with relatively few branches

(ii) there is little undergrowth to hamper the work.

In Sweden, dense coniferous forest covers more than half the country. Rivers provide natural waterways for log transport and H.E.P. for the sawmills. The timber is used for
(i) building
(ii) wood pulp – used for paints, varnishes and industrial alcohol
(iii) paper
(iv) fuel – because Sweden has no coal
(v) matches
(vi) furniture.
Sweden concentrates increasingly on manufactured wood products because they are more valuable than the raw material. Wood and wood products account for 20 per cent of Swedish exports.

Together with the benefits of forestry, Sweden has the disadvantages of pollution:
water – fouled with bleach from paper factories
air – fouled with poisonous fumes.

Rain forests
Rain forests are difficult to exploit because
(i) different species (e.g. mahogany) are widely scattered
(ii) trunks and roots are often interlocked
(iii) undergrowth is dense
(iv) almost all the rain forests are in less developed countries where transport conditions are bad and electricity is not easily available.
Ghana's rain forests are in the south-west. The trees – obeche, sapele and mahogany – are cut by hand-saw or axe and dragged by crane to be loaded on to railway wagons or lorries. Most of Ghana's logs are exported unprocessed but the sawmill industry – catering for the domestic market – is the largest industry in Ghana. Furniture is made from Ghanaian wood, but hardwoods are not good for papermaking.

Water

You should by now realize the importance of water
as a raw material (p. 107)
as a coolant (p. 97)
as a source of power (p. 138).
In southern Africa, shortage of water is holding up further industrial development.

World Industry

Fig. 136 shows the largest industrial regions. You should note that the belt from the centre of the U.S.A. to the Ural mountains of the U.S.S.R. accounts for 80 per cent of the world's manufacturing by value. Note also the few industrial areas of the developing world – underlined in fig. 136. As well as knowing the major industrial areas of your chosen region in detail you should know where the other industrial areas of the world are to be found. Look for sections on industry in Chapter 8. This is particularly important for J.M.B. candidates.

World Agriculture

You must be absolutely clear about the meaning of the expressions used. (see also pp. 121–2)

Commercial agriculture is the production of food for sale. It can be extensive or intensive. Some different types are:

(i) **intensive pastoralism** – dairying in Brittany; Upper Michigan

(ii) **extensive pastoralism** – beef in Texas

(iii) **intensive arable** – market gardening: New Jersey, Flanders

(iv) **extensive arable** – wheat on Prairies and Great Plains

(v) **intensive mixed** – 'Corn Belt', U.S.A. (maize)

(vi) **plantation agriculture** is a common form of commercial farming in the developing world, e.g. coffee in Brazil, tea in Sri Lanka.

Subsistence agriculture is farming for consumption by the farmer's family.

(vii) **Shifting cultivation** – when a farmer abandons land after a few years and leaves it to revert to bush – is increasingly rare

(viii) **nomadic herding** – practised, for example, by the Bedouin people of North Africa; also increasingly rare.

Sometimes a subsistence farmer also grows some **cash crops** for a little income: here commercial and subsistence farming overlap.

You will be expected to be able to locate areas where each of these farming types exists, as well as production areas of wheat, maize, rice,

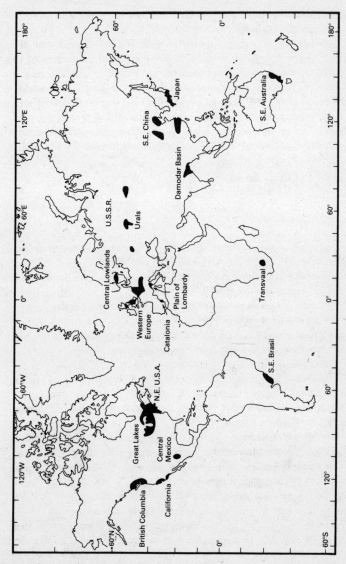

Fig. 136 *World industrial regions*

groundnuts, sugar beet, cocoa, coffee, tea, **mediterranean produce** (citrus fruits, wine, olive oil), flax, jute and cotton (fig. 137).

Crop	Conditions	Notes	Example
Wheat	rainfall – 600–750 mm, summers warm or very warm soils – fertile chernozem relief – flat, to aid mechanization	spring wheat (sown in spring) needs warm, rainy summers winter wheat grown	Canadian Prairies U.S.A.: Great Plains
Maize	summer rain, sunshine, frost-free six months	irrigation often used	U.S.A.: Mid-West
Rice	temperatures above 20°C; 750 mm of water during growing season relief – flat plains or terraces to be flooded	staple food of half the world	India
Groundnuts	temperatures above 20°C; 500 mm of rain	used more for oil than for peanuts	N. Nigeria
Cocoa	rainfall – 1,200 mm, evenly distributed deep, well-drained soils		Southern Ghana
Coffee	rainfall – 750 mm (summer max.), temperatures over 20°C well-drained soils – slopes preferred		Uganda
Mediterranean produce	hot summers, mild winters, rainfall moderate, winter max.	irrigation usually necessary in summer	Southern Spain

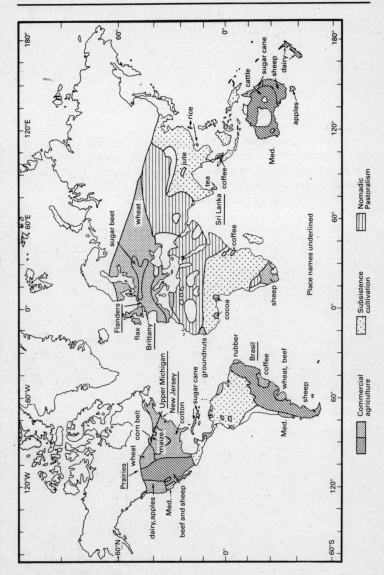

Fig. 137 *World agriculture*

7. Seasons and Time

Questions on this topic are very direct, only requiring a basic understanding of a few principles, rather than the memorizing of strings of facts. The key to understanding this section is that THE EARTH IS TILTED IN RELATION TO ITS ORBIT AROUND THE SUN (fig. 138). In consequence, at any one moment, the sun's rays strike the earth at different angles in different places (fig. 139). The **altitude** of the sun (the angle of the incoming rays at noon) is 20° at A and 90° (overhead) at B.

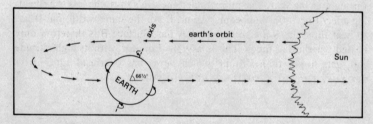

Fig. 138

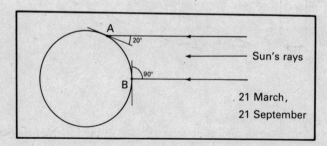

Fig. 139

Latitude

Latitude is calculated by subtracting the noon altitude of the sun from 90°:

latitude of A = 90°–20° = 70°north

latitude of B = 90°–90° = 0° = Equator.

A line joining points of the same latitude forms a circle parallel to the Equator: a **parallel of latitude**. There are five important parallels:

(i) *Equator* – where day and night are always twelve hours long

(ii) *Tropic of Cancer* (23½°N) – sun overhead on 21 June

(iii) *Tropic of Capricorn* (23½°S) – sun overhead on 21 December

(iv) *Arctic Circle* (66½°N) – twenty-four hours' continuous darkness on 21 December

(v) *Antarctic Circle* (66½°S) – twenty-four hours' continuous darkness on 21 June.

From (ii) and (iii) above, you can see that the sun's altitude varies throughout the year. The more angled the sun's rays, the less their heating effect (fig. 140). 'Sunbeams' A and B are the same width, but B has to heat three times the surface that A has. Surface B is therefore only heated one third as much. This is why the Equator, with its high altitude sun, gets hotter than Britain, which never has overhead sun – NOT BECAUSE THE EQUATOR IS NEARER THE SUN.

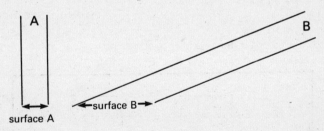

Fig. 140

Day and night; the seasons

As we lean towards and away from the sun so the length of time we are in shadow (night) and in sunlight (day) varies (fig. 141). You can see that, along the line of latitude 40°N:

on 21 June day is twice as long as night (**summer solstice**)
on 21 December night is twice as long as day (**winter solstice**).

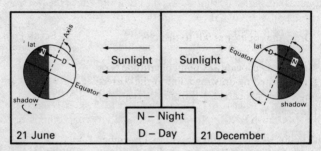

Fig. 141

The change from the long days of June to the long nights of December
is gradual. Half-way through each period of change – on 21 September
and 21 March – day and night are of equal length; these are the **equinoxes**.

Add to the changing lengths of day the changing angles of the sun
and you will understand why temperatures vary with the seasons.

Look again at fig. 141 and you will see the equal lengths of daytime
at the Equator in June and December.

*Q. 1 For the city of Sao Paulo 23°30'S, 47°W, state and explain, with
the aid of diagrams, the elevation of the noonday sun on 21 December*

[LON A, part question]

1. With a compass draw a circle to represent
   the earth

2. Lightly draw a horizontal line across the
   middle

3. Draw the axis, through the centre, at $66\frac{1}{2}°$
   from the horizontal

4. Label North Pole and South Pole; rub out
   the horizontal

5. Draw the equator at 90° to the axis

6. Now you can locate any point of latitude N or
   S as required by drawing a line at the correct
   angle – in this case 23½°S from the centre of
   the equator

7. Finally locate the sun's rays. In December
   they will fall on the side where the south pole
   will be in the sunlight. In June, they will fall
   onto the "North Pole side"

Fig. 142

If you practise steps 1–5 this construction will become easy, but
accuracy is important.

Longitude

Lines of longitude run at right angles to the parallels of latitude. 0° runs
through Greenwich because it was there that the system was perfected.
Together with lines of latitude, longitudinal lines make up a grid reference
system.

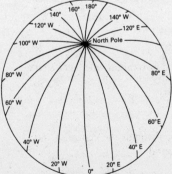

Fig. 143

Longitude and time

All points along the same line of longitude have the same time. The earth rotates through

   $360°$ in 24 hours

   $15°$ in 1 hour

   $1°$ in 4 minutes.

Time to the west is earlier, to the east is later. Thus time along the line of longitude $1°$ west of Greenwich will be four minutes earlier than at Greenwich.

If you know the difference in longitude between two places, you will be able to calculate the time difference. For example, if it is noon at $10°W$, what is the time at $22°W$?

Difference in longitude $= 22°–10° = 12° \times 4 = 48$ minutes earlier i.e. it is 11.12 a.m. Be careful when you calculate differences where one line is east of Greenwich and one line is west (fig. 144).

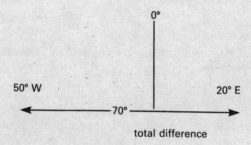

Fig. 144

Thus, if it is 11 a.m. at $20°E$, what time is it at $50°W$?

   Total difference $= 70°$

   $70 \times 4 = 280$ minutes $= 4$ hours 40 minutes

   Time at $50°W$ is 4 hours 40 minutes earlier, i.e. 6.20 a.m.

Sometimes a rhyme can help your memory.

> *Across the longitudes times range*
> *For each degree, four minutes change*
> *New Yorkers eat their breakfast*
> *When it's lunch in London town*
> *But it's supper in Australia:*
> *Because the world spins round.*

The International Date Line is the name given to the 180° **meridian** (line of longitude). Going from, for example, Japan to California, you lose a day. Going the other way, you have one day twice.

Q. 2 (a) Study carefully [fig. 145], which shows four positions of the Earth in its revolution around the Sun.

(i) In the spaces provided, insert the correct dates of the solstices and equinoxes which are labelled A to D.

(ii) On the diagram shade in those parts of the Earth which will be in darkness at the four positions shown. [8]

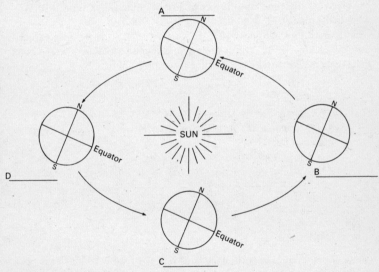

Fig. 145

Be prepared for diagrams shown from unusual angles:

Q. 2 (b) [Fig. 146 overleaf] shows selected lines of latitude and longitude centred on the South Pole.

(i) Clearly label the International Date Line;

(ii) In the spaces provided state the longitude of L_1 and of L_2 and the local time at T_1 and at T_2. [5]

(c) (i) Calculate the length in kilometres of the Greenwich Meridian shown on fig. 146, given that 1 degree of latitude = 112 km.

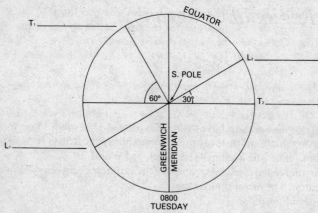

Fig. 146

> (*ii*) *Explain one of the following:*
>   (*1*) *the changes in time which are necessary when crossing the International Date Line;*
>   (*2*) *how latitude differs from longitude;*
>   (*3*) *Great Circles, and why they are important for air routes.*

[10]
[LON A]

Great Circles

Any line dividing the earth exactly in half is called a **Great Circle**. The shortest route between any two points on the earth lies along the Great Circle which passes through them, but when drawn on a flat map, Great Circle routes do not look like short cuts. These routes are most important for aeroplanes since there are no relief obstacles to interrupt the Great Circle route.

8. Regional Geography

For every syllabus, you have to be able to read O.S. maps. You must also answer questions on the regional geography of the British Isles. (This may come under the heading 'The Developed World'.)

It is in the second regional section that the variations between syllabuses are greatest. The major options are:
1. Western Europe
2. North America
3. 'World Population Zones'
4. 'World Problems'
5. Aspects of World Geography
   (i) The developed world (*but not the U.K.*);
   (ii) The less developed world
6. The Developing World (Cambridge syllabus).

The Joint Matriculation Board's syllabus A has no second regional section. Accordingly you will need to find more detail and more named examples to supplement the information in Chapter 6 in the sections on 'Relief', 'Climates', 'Vegetation', 'Resources', 'Industry' and 'Agriculture' – notably the production of major food crops.

Western Europe

Many of the themes in the geography of Western Europe are very similar to those on the British Isles. Consequently, the types of question are also alike.

In this section you will frequently be asked to re-read the section on the subject in the chapters on Britain (Chapters 2 to 4) before reading how the same principles affect Western Europe.

Climate

Check the factors affecting climate (p. 41). Note the following variations.

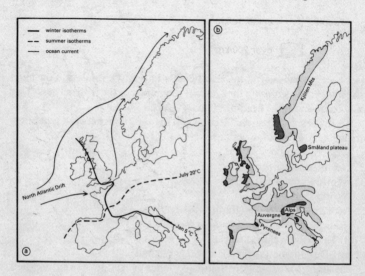

Fig. 147

1. *Altitude.* The Scandinavian mountains and the Alps rise well over 2,000 m; the effect on temperatures is therefore greater than in Britain.
2. *Latitude.* Europe extends further south than Britain and temperatures in, for example, the south of France are correspondingly higher.
3. *Depressions* only affect southern Europe in winter.

Types of question
1. Account for and suggest locations for climate graphs.
2. Explain causes of, and solutions to, problems of drought and flooding.

Population

Check the negative and positive factors affecting population (p. 81). You should be able to identify the sparsely populated regions of Western Europe (fig. 148), A–E, using an atlas.

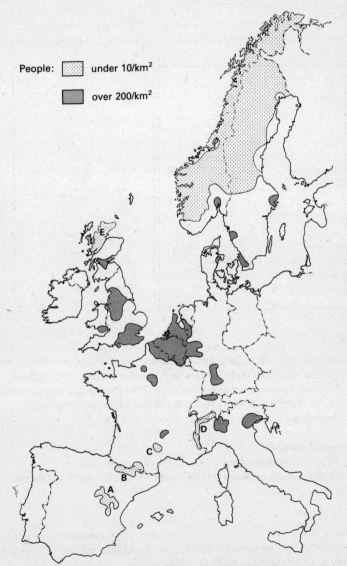

People: ▫ under 10/km²

      ▪ over 200/km²

Fig. 148 *Very dense and very sparse population in western Europe*

Population distribution in Scandinavia

Since Scandinavia shows such extremes of population density, it is a commonly examined area.

Revision Exercise 1

1. Locate, describe and explain the:
   (a) very sparsely populated
   (b) densely populated
   regions of **Scandinavia** (Norway, Sweden, Denmark).
2. Construct a map of Scandinavia showing land over 1,000 m
   Norwegian Sea (mark in North Atlantic Drift (p. 202) and the Baltic (write in 'frozen in winter')
   fertile areas – Oslo Lowlands, Swedish Lakes Lowland, Skåne
   minerals and metal-working centres
   coniferous forest
   fishing area: Lofoten Islands
   Ekofisk oil- and gas-field
   Esbjerg, Alborg, Bolense, Copenhagen
   Malmö, Göteborg, Norrköping, Uppsala, Stockholm, Luteå
   Oslo, Drammen, Stavanger, Bergen, Hammerfest, Harstad, Narvik
3. Recent developments
   In 1960, oil and natural gas were discovered under Norwegian waters. Pipelines cannot be constructed to the Norwegian coast, because of the Norwegian Trench. Petrochemicals at Rafres and oil-refining at Harstad have increased population in northern Norway.
4. You should be able to draw sketch maps to show the site and situation of Oslo, Göteborg, Stockholm and Copenhagen. In each case, wherever relevant, draw and name
      uplands (over 200 m)
      sea coast
      major communications
      industrial or agricultural hinterlands.

You can work out such a map from your atlas.

Conurbations

The major European conurbation is Randstad Holland (fig. 149).
Growth of Randstad Holland resulted from

1. Rural depopulation due to decline in agricultural workforce led to urbanization
2. Processing industries based on imports through Ijmuiden, Europoort, Rotterdam
3. Petrochemicals – Overijssel gas-field
4. Market-based consumer industry
5. Tertiary industry.

Fig. 149 *Randstad Holland*

Randstad Holland has the problems of conurbations listed on p. 80. Many of the solutions resemble those discussed on p. 134, but the Dutch also specialize in land reclamation to increase the area available for development, p. 136.

Coal

Problems of European coalfields are listed on p. 95. This topic is not often examined, except in connection with heavy industry (p. 94).

Power

Power is supplied in Western Europe by the various methods noted on pp. 97–9.

Revision Exercise 2

Explain the distribution of
  (a) H.E.P. plants in Norway and Sweden
  (b) power stations in the Netherlands with regard to the chief physical
      and human locational factors (fig. 150). Note that the type of power
      station in Norway and Sweden is quite different from that in the
      Netherlands.

(a)   (b)

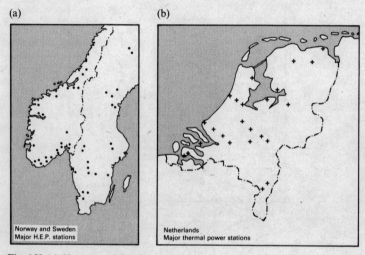

Fig. 150 (*a*) *Norway and Sweden – main H.E.P. stations;* (*b*) *Netherlands – main thermal power stations*

Natural gas

The U.K., the Netherlands, Norway, France and Italy have benefited
most from their own resources of natural gas. Gas is normally piped
to pre-existing industrial areas, so its importance has been as a source
of
  (i) cheap power
  (ii) revenue, e.g. the Netherlands sell gas to West Germany
  (iii) raw materials.

Natural gas in Holland has been used to develop industry in the northern Netherlands – an area of high unemployment. Delfzijl has an aluminium smelter and chemical and fertilizer ports.

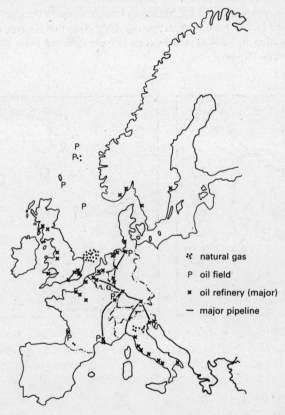

Fig. 151

Oil

Oil is also highly transportable. Fig. 151 shows that the major refineries are located on deep-water harbours or along the pipelines which link the oil ports with inland industrial areas.

Oil is a major raw material in the chemical industry (fig. 152), but other raw materials are used (see north Cheshire, p. 109).

Fig. 152

Note that, for southern Germany, the oil ports of Marseilles and Genoa are as close – as the crow flies – as those on the North Sea coast. The **C.E.L.** has had to be built across the Alps, over uplands of 2,000 m O.D. so that low temperatures have added to problems of gradients.

Industrial development

You may be asked the range of industries in one country, in which case you should base your explanations on the 'Five Factors': power, raw materials, markets, labour, transport (p. 106). Information may be given you in the form of a **pie diagram**.

Q. 1 Study [fig. 153], which shows the proportion of workers employed in groups of industries in Sweden. Explain this pattern.

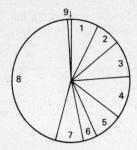

1 Food, beverages
2 Textiles
3 Wood, wood products
4 Paper, paper products, printing
5 Chemicals, chemical products
6 Non-metallic products
7 Basic metals
8 Manufactured metal products, machinery
9 Other

Fig. 153

To read a pie diagram you must have a protractor. To calculate the proportion of Swedish workers employed in, for example, group 8, measure the angle at the centre of the segment and divide by 360°, i.e. $\frac{160}{360}$; cancelling gives $\frac{4}{9}$, i.e. just under half. Use estimates rather than being scrupulously accurate.

More often, a specific industry is examined, most notably iron and steel.

Western Europe: Iron and Steel

Locational factors. (1) *In the nineteenth century* coking coal was vital: ten tonnes of it were needed to produce one tonne of iron. Because of transport costs, the industry was tied to coalfields, e.g. the Ruhr (Germany), Sambre–Meuse (Belgium and France). Iron ore was moved along rivers and specially constructed canals. (2) *Twentieth-century developments* have been affected by

> **beneficiation** – partly refining the ore at source
> improved technology – less coal is needed
> exhaustion of nineteenth-century coalfields
> the creation of the European Coal and Steel Community in 1952, which
> > tried to plan the industry efficiently.

New steelworks are planned on **tidewater** sites, to use cheap sea transport, and in areas where flat land is available, e.g., Ijmuiden (Netherlands), Dunkirk (France).

Often the more inefficient inland sites are kept in production because of the amount of capital and skill that have been invested in them and because of the problems of unemployment resulting from steelworks closures.

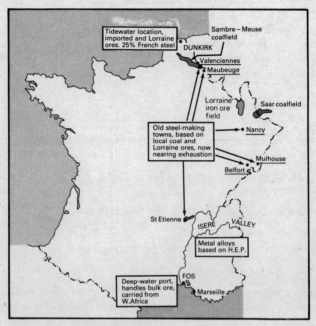

Tidewater location, imported and Lorraine ores. 25% French steel

Sambre – Meuse coalfield

DUNKIRK

Valenciennes

Maubeuge

Lorraine iron ore field

Saar coalfield

Old steel-making towns, based on local coal and Lorraine ores, now nearing exhaustion

→ • Nancy

Mulhouse

Belfort

St Etienne

ISERE VALLEY

Metal alloys based on H.E.P.

Deep-water port, handles bulk ore, carried from W.Africa

FOS

Marseille

Old-established sites <u>underlined</u>
New sites in CAPITALS

Fig. 154

The Belgian section of the Sambre–Meuse coalfield also developed iron and steel, e.g. at Liège. Like the French coalfield locations, there have been problems of exhausted coal reserves, congested sites and pollution. The newest Belgian steelworks is near the coast at Zelsate, another tidewater, flat-land location.

The West European and North American steel industries are suffering from:

over-production – due to the world recession
competition – from Eastern Europe and Japan.

Transport

Ports

You should be able to produce sketch maps to show
  (i) the site
  (ii) the hinterland
of Marseilles, Fos, Rotterdam and Europoort, Hamburg, Copenhagen, Stockholm and Göteborg. The maps of Fos (fig. 155) will serve as an example.

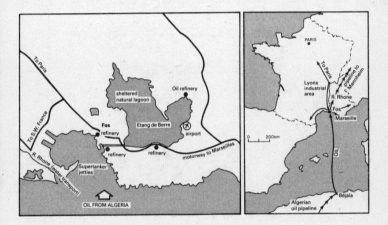

Fig. 155 *Fos – site and hinterland*

Rivers

River transport is much more important in Europe than in Britain. Fig. 156 shows a **flow-line graph** of German waterways (rivers and canals). You are meant to *use* the graph. No scale is given but you can compare river traffic, e.g. that on the Rhine with that on the Danube, by using proportions: the traffic on the Rhine at the Dutch frontier is twenty-one times that on the Danube at the Austrian frontier. (Use your ruler to measure width of lines.)

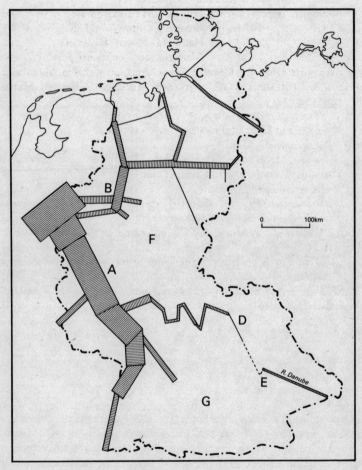

Fig. 156 *West Germany – main navigable waterways*

Revision Exercise 3

1. Draw an outline map of West Germany and the Netherlands.
2. Mark
   rivers: Rhine, Main, Neckar, Saar, Danube
   canals: Mittelland, Dortmund–Ems, Elbe
   ports with 'import' arrows: Emden – 'Swedish iron ore'
              Bremen – 'cotton from U.S.A.'
              Hamburg – 'Swedish iron ore'
              Bremerhaven – 'container port'.
   river ports: Duisburg, Cologne, Frankfurt, Ludwigshafen, Mannheim
   industries: CHEMICALS at Leverkusen, Rheinport, Hoechst, Mannheim (fig. 152)
       TEXTILES, Ruhr valley
       ENGINEERING, Ruhr valley
       COAL, Ruhr, Saar
       WINE, Mosel valley, Rhine Gorge.
   You should now be able to attempt Question 2.

Q. 2 with reference to [fig. 156]
   (a) Give an account of the Rhine (A) as a navigable waterway, mention-
   ing the main commodities transported   [6]
   (b) What is the importance of the waterways in the area between B
   and C?   [6]
   (c) What may be the main consequences of the completion of the deep
   canal linking D and E?   [3]
   (d) Why are there no important waterways in the areas marked F
   and G?   [3]

European rivers are also more important for power and irrigation: the Rhône provides both (fig. 157).

You will not be expected to draw maps of roads or waterways in West European countries, but you should be able to link the transport networks of at least one country with its economic development.

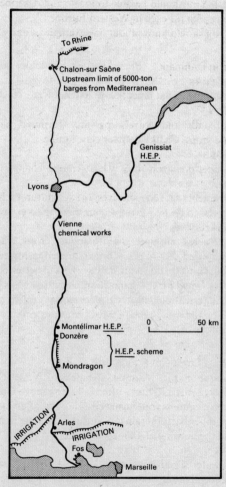

Fig. 157 *The Rhône*

Agriculture

The table on pp. 122–3 summarized the conditions required for production of beef and dairy cattle, wheat and barley, root crops (especially sugar beet), orchard crops, market gardening and early fruit and vegetables (**premium crops**). You should be able to locate and describe at least one major producing area for each in Western Europe.

In addition you should know about one example of dairy farming.

Dairy produce in Denmark
PHYSICAL FACTORS
Relief. Gently sloping or flat land, below 200 m.
Climate. See fig. 147.
Soils. Infertile sands and gravels: pasture. Terminal moraine area: pasture, some crops. Chalky boulder clay: arable.
HUMAN FACTORS
Historical. Competition from U.S. wheat caused change from cereals to dairying.
Social. Establishment of (i) cooperatives; (ii) agricultural schools.
Markets. Exports to U.K., E.E.C. countries; market gardening for domestic markets.
Typical farm 52 acres, all given over to fodder crops, cattle and pigs permanently housed, fed on grass 'bricks' and concentrates.
Problems Labour shortage due to drift of workers to better-paid industry; some farms too small for mechanization; increased costs of milk production; change in eating habits – less demand for fats.
You should also study one type of Mediterranean product, e.g. vines, olives, citrus fruits.

Problems in West European Agriculture
1. Excess precipitation. Fig. 147 shows the areas of Western Europe with over 1,500 mm of precipitation – often falling as snow. One solution to lack of winter pasture is **transhumance** – moving of animals and their herdsmen to upper pastures (**alp** in Switzerland, **saeter** in Norway) – in summer, and the housing of animals in winter.
2. Drainage, see p. 136.
3. Shortage of water. The Rhône irrigation works (p. 157) serve the Midi and Languedoc areas of France. For details of Mediterranean climate and vegetation, see pp. 172 and 183.

Individual countries

You should be able to write a **geographical account** of two countries. This would include Relief and Drainage, Climate, Vegetation (if an important natural resource), Natural Resources, Agriculture, Industry, Population Distribution.

Belgium and Switzerland would be good choices because they are small, have a fairly simple outline, and contrast strongly with each other in many respects. Be prepared to draw a map of the **natural regions** of each country.

North America: Canada and the U.S.A.

This region is part of the developed world and thus has many similarities with Britain. Consequently, you will be referred to themes noted in the British Isles section and shown briefly how to apply them to North America. We will only look in detail at developments not found in the U.K. or Western Europe.

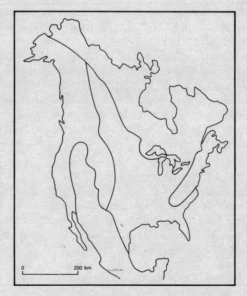

Fig. 158

Relief and Climate

Revision Exercise 4

Make a large copy of fig. 158. Using an atlas, mark on it:

    structure, using fig. 159.

    mountains: Appalachian, Rockies, Sierra Nevada, Cascades

    rivers: St Lawrence, Saskatchewan, Mississippi, Ohio, Hudson, Colorado, San Joaquin, San Jose, Columbia

    provinces: Quebec, Labrador, Ontario, Manitoba, Saskatchewan, Alberta, British Columbia

    states: California, Florida, Pennsylvania, New York State, Alaska, Utah

    regions: New England.

Fig. 159 shows major outlines of climate. Check the factors affecting climate (pp. 175–6).

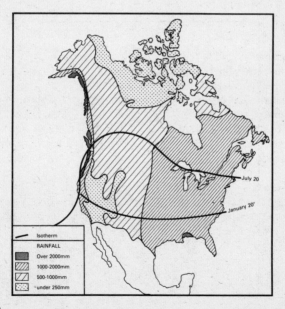

Fig. 159

Mark in
> the California Current (fig. 131)
> the Gulf Stream
> incoming westerly winds from the Pacific
> (i) RELIEF RAIN along the West Coast (p. 34)
> (ii) CONVECTIONAL RAIN over Prairies and Great Plains (p. 34)
> (iii) RAIN SHADOW over Nevada, Utah, Arizona
> (iv) DEPRESSIONS over the Great Lakes (pp. 32–3)
> (v) HURRICANES along Gulf and East Coast (pp. 176–8).

Satisfy yourself that you know how these five weather features occur; page references are given.

You might answer Question 3 using Alaska and Utah (p. 135) as examples.

Q. 3 In North America, select
> *(i) one area with a hostile climate*
> *(ii) one area where relief hinders development*
> *For each area, explain how man combats these handicaps.*   [OXF]

Population

Distribution

You should know the factors which explain
> low population density, e.g. Alaska, Labrador
> high population density, e.g. California, New York State, Southern Ontario.

Take physical factors first, in the order in ReDisCoVeriNg (pp. 152–3), then human – Transport, Markets and Labour – and show how they affect economic activity.

Conurbations

The major example in the U.S.A. is **Megalopolis** (fig. 160). You should know: the reasons for the development of Megalopolis, the problems of the conurbation (p. 88), and some solutions (p. 134).

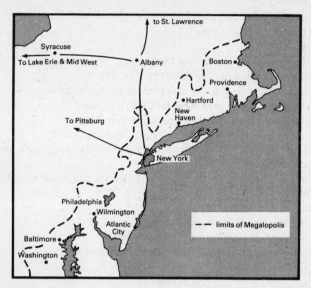

Fig. 160 *Megalopolis*

Cities

Practise drawing *site* and *situation* maps (the maps of Fos on p. 211 will demonstrate the difference between these two) for

    Montreal, Ottawa (Canada)

    Boston, New York, Philadelphia (East Coast U.S.A.)

    Chicago, Houston

    Seattle, Los Angeles (West Coast U.S.A.)

Areas of population growth

    Alaska (p. 135)

    Alberta

    Arizona

are all increasing in population. Why?

Urban morphology

The easiest example to learn is New York. (New York is NOT the capital of the U.S.A.)

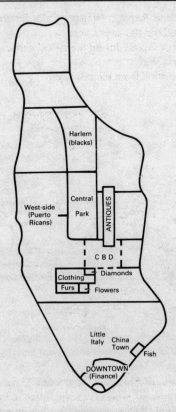

Fig. 161 *Central New York City: Morphology*

Transport

Water

The Great Lakes and St Lawrence Seaway (fig. 162) constitute a unique waterway system over 2,000 miles long. Close to it occur the three main raw materials for steel-making: iron ore, coal and limestone; and the major agricultural areas of the Canadian Prairies, the American 'Corn (Maize) Belt' and Great Plains. This was the incentive to link the five Great Lakes by canals and to deepen the St Lawrence river. The system developed over 130 years. Major problems were physical: Lachine Rapids, Niagara Falls freezing – the seaway is closed for 100 days a year

economic: lack of capital during the 1930s' depression and the Second World War

opposition from the railways.

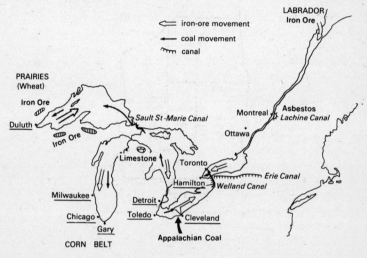

Fig. 162 *The Great Lakes*

Fifty-nine per cent of Canadian freight and 53 per cent of U.S. freight travels by water transport, which is cheap, suited to bulk cargo, but slow. The U.S. Mississippi–Missouri system is also important. Other rivers, like the Mackenzie, have not developed because of lack of economic incentive.

Other functions of the Great Lakes are

(i) provision of water
   as industrial raw material
   as coolant
   for domestic use

(ii) generation of H.E.P. at, for example, Ottawa, Beauharnais

(iii) recreation: there are holiday resorts on the Upper Lakes.

Railways

Fig. 101 shows that railways are very competitive for distances over 400 km and yet rail traffic in North America has declined because of competition from

pipelines

waterways

air transport – particularly for passenger traffic.

To halt the decline, railway companies have introduced

containers

piggy-back transport (truck bodies carried by rail can be hitched to motors at destinations).

Coal, minerals and fuel

Re-read pp. 93–103. The decline of coal, the increased use of oil, the location of oil refineries and oilfield industrial developments, the requirements for H.E.P. and the siting of nuclear power stations in North America are all controlled by similar factors to those operating in Britain and Western Europe. You must provide the examples. Most frequently examined are:

coal: the Appalachians

oil: the Gulf

H.E.P.: learn a multi-purpose river project (p. 138) e.g. the Colorado.

Industry

Learn how and why the locational requirements of the steel industry have changed. Find examples from the U.S.A. to illustrate this.

Recent developments in American industry

In the last thirty years, American industry has changed because

1. The use of new materials, like plastics, has changed the raw materials required. Thus new locations have become important. Tidewater oil refineries have replaced coalfields as growth areas.
2. The rising standard of living has increased demand for consumer goods: light industry, less tied to raw materials.
3. Electricity and natural gas are easily transportable. Manufacturers can locate in pleasant areas like Arizona, away from the polluted, congested coalfields.
4. Tertiary industry has grown in importance. Atlanta (Georgia) has developed rapidly since 1960 as a centre of

warehousing
distribution of goods to smaller centres
offices for major manufacturers.
All this is based on its excellent **communications** (transport).

Agriculture

Re-read the best conditions for various types of agriculture on pp. 122–3 and 192 and provide sample North American locations for each. In addition, you should know about cotton, tobacco and sugar cane. Complete this table:

	Locational Factors	*Sample Areas*
Cotton		
Tobacco		
Sugar cane		

You should also know about the increasing importance of soya beans.

Other aspects of agriculture are best covered in **small regional studies**. You should be able to write a geographical account (see p. 216) of California; Florida; the Prairie Provinces.

World Population Zones

These are examined by Joint Matriculation Board's syllabus B. Listed below are the requirements of this section of the syllabus, with references to the relevant pages in this book:
1. World distribution of population (pp. 150–54)
2. Almost uninhabited areas (p. 152)
3. Sparsely populated areas, including areas of economic importance but small size, like the Rand area of the Republic of South Africa (fig. 163 overleaf).

Revision Exercise 5

Using fig. 163, write an account of mining and industry on the Rand under the following headings:
   mining – raw materials, chief locations
   iron and steel – power, raw materials, transport.

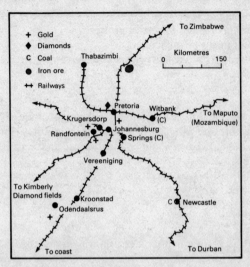

Fig. 163 *The 'Rand' area of South Africa*

Look up 'Labour' and 'Markets' (gold comprises 48 per cent of the exports from Republic of South Africa).
4. Densely populated areas
   (i) an **occidental** (western) area, e.g. Western Europe (pp. 201–15), but not the U.K.
   (ii) an **oriental** (eastern) area (pp. 224–7).

Densely populated oriental areas

Fig. 164 shows areas of the Far East with more than 200 inhabitants/km². You will need to know
   (i) reasons for high population density (see pp. 146–7)
   (ii) rural economies
   (iii) areas of food deficiency
   (iv) urbanization
   (v) industrialization.

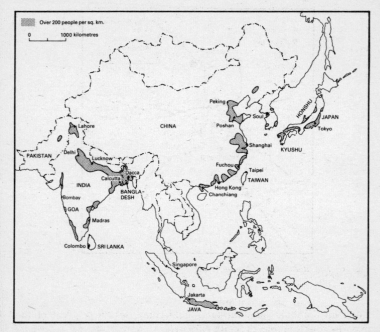

Fig. 164 *The Far East – regions of dense population*

Rural economy

Rural economy means the way the country people live. Over much of Asia, *padi* (rice) farming is the basic farming method.

The table overleaf shows the year's activities for one farmer in Bihar, India. He has four fields, total area 0.6 acres, scattered throughout the village. Note how closely the farmer's activities are tied to rainfall.

Month	Rainfall (mm)	Tasks
Jan.	10	irrigating, weeding
Feb.	31	harvest *rabi** crops
March	36	repair irrigation ditches
April	43	ploughing
May	140	sowing: rice, maize, *kharif*† crops
June	297	manuring
July	325	weeding
Aug.	328	weeding
Sept.	252	harvest *kharif*† crops, ploughing
Oct.	114	sowing: wheat, barely, pulses, *rabi** crops
Nov.	20	irrigating, manuring
Dec.	5	irrigating, weeding, deepening wells

* spring crops
† autumn crops

Urbanization

Urbanization has been discussed on pp. 157–9 and the major cities of South-East Asia are shown in fig. 164. Hong Kong has tried to overcome problems of urbanization by building

(i) flats – a family of four would have one room, washroom and toilet

(ii) 'flatted' factories – one factory per floor – to provide employment

(iii) local welfare centres, nurseries, schools, libraries, clinics to cut down travel into the city centre.

Industrialization

Fig. 165 shows the Damodar Valley, the major industrial area in S.E. Asia. Use p. 106 to remind you of the 'Five Factors' influencing industry. Then write an account of the advantages of the Damodar Valley for industrial growth, based on fig. 165. For labour and markets, refer to fig. 164. Further advantages are the low cost of Indian labour.

Problems of industrialization in India include:

1. Lack of capital: 70 per cent of the money to develop the Damodar Valley came from loans. India has had to ask for aid just to repay the interest on the loans.

2. Increasing population, which offsets the gains made (p. 147).

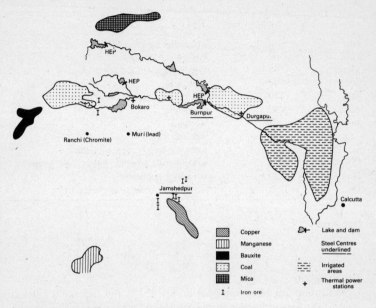

Fig. 165 *The Damodar Valley*

238   *Passnotes:* **Geography**